Troubleshooting Vacuum Systems

Scrivener Publishing
100 Cummings Center, Suite 541J
Beverly, MA 01915-6106

Publishers at Scrivener
Martin Scrivener (martin@scrivenerpublishing.com)
Phillip Carmical (pcarmical@scrivenerpublishing.com)

Troubleshooting Vacuum Systems

Steam Turbine Surface Condensers and Refinery Vacuum Towers

Authored by

Norman P. Lieberman

Scrivener Publishing

WILEY

Co-published by John Wiley & Sons, Inc. Hoboken, New Jersey, and Scrivener Publishing LLC, Salem, Massachusetts.
Published simultaneously in Canada.

For general information on our other products and services or for technical support, please contact our Customer Care Department within the United States at (800) 762-2974, outside the United States at (317) 572-3993 or fax (317) 572-4002.

Wiley also publishes its books in a variety of electronic formats. Some content that appears in print may not be available in electronic formats. For more information about Wiley products, visit our web site at www.wiley.com.

For more information about Scrivener products please visit www.scrivenerpublishing.com.

Cover design by Kris Hackerott

Cover image courtesy of Roy Williams

Library of Congress Cataloging-in-Publication Data:

ISBN 978-1-118-29034-7

MWEP451587090426

To Liz – Wife, friend, and partner in life. She has climbed many a vacuum tower structure with me. Infrared temperature sensing gun in hand, Liz has provided innumerable surveys for steam ejector and inter-stage condenser surface temperatures for vacuum systems at the most remote corners and hostile environments of our little home planet.

Contents

Preface

Challenging Human Intellect

You might think that the author of a book pertaining to trouble-shooting vacuum systems would be super successful in resolving such problems. Especially considering I've had 47 years of experience and practice on literally a thousand such systems, *and* considering that I am, at least in my own mind, really smart. But you would be very much mistaken.

I rarely resolve all the problems of vacuum systems on my first try. Often, I'll require several retrofit attempts to eliminate all the malfunctions. Not infrequently, I'll miss the point completely and only years later will someone else discover the underlying problem.

Process engineering is one of the most difficult activities that mankind has selected in our desire to dominate the universe. And, the performance of multi-stage ejector-condenser systems is clearly one of the most complex features of process engineering. The performance of such systems lies at the outer edge of average human comprehension. I guess that if one could engage the services of the very top intellects who have fanatically dedicated and devoted 100% of their time to vacuum systems, then surface condensers and vacuum towers could consistently be operated at design pressures.

But in reality, the vast majority of such systems are operated at some fraction of their design efficiency. Thus, there is, and will continue to be, a huge economic incentive in refineries, petrochemical plants, and power generation plants to correct vacuum system malfunctions.

So many of my clients operate with inefficient vacuum systems for so long, that they consider their current operations normal. They have no conception of the increased product yields or the energy savings that would result if they could only develop the design vacuum.

Often, I find my inability to identify a vacuum problem to be depressing. I'll be all alone, on the 8th landing, in the cold and damp of a dark night, listening to a great jet surging. The loneliness and isolation of the situation just makes me colder.

"Now what?!" I'll think. I'll be at the limit of my mental capacity to grasp the malfunction. But then, the defining characteristic of humankind comes to my aid. Not intelligence, but determination.

Need Help?

It is my intention in writing this book to contribute to the reader's ability to improve the performance of their vacuum systems. This has got to be a "Hand's-on" activity. However, often it may help you to discuss your observations with me. I never charge for such consultations. You can reach me at:

1-504-887-7714 (phone)
1-504-456-1835 (fax)
norm@lieberman-eng.com
www.lieberman-eng.com

Introduction

Time and Determination

Vacuum system malfunctions are treated in the process industry, and especially in my universe of petroleum refineries, differently than other sorts of process equipment malfunctions. Problems with pumps, compressors, fired heaters, and distillation towers are eventually resolved and corrected. Problems with vacuum systems are only too often never fixed or even recognized, meaning my clients just become accustomed to running with a bad vacuum.

A young engineer assigned to improve the vacuum in a lube oil asphalt, or a coker feed vacuum tower, will typically tread down the following path of frustration:

- Consult texts on vacuum equipment. Usually a waste of time. These texts describe how the vacuum equipment is supposed to work and not how it reacts to malfunctions such as wet steam, defective condenser seal strips, and eroded steam nozzles.
- Consult the equipment vendor. Totally a waste of time. These fellows do not understand how their jets react with the process itself and the associated equipment.
- Consult with a senior engineer. No help here. These gentlemen and ladies are always on their way to an important meeting and are too busy to work on such a humble task as ejector malfunctions.

So, the first step in troubleshooting ejectors and vacuum systems is to accept the fact that you'll have to do it by yourself. And how about reading my book that you've just purchased—the very text that you have in your hands right now? Well, if you will really read this book in its entirety, you will have a growing sense of desperation and depression. I know. I've been there a hundred times. It's all so terribly complex.

After all, if no one has solved this problem in the past 20 years, why would anyone expect you to resolve it? Perhaps it's best to make some superficial, long term recommendations and get on with your life. Here are a few such recommendations that you might suggest to your upper management and supervisors:

1. During the next unit turnaround, disassemble the ejector and ship the internals back to the manufacturer for testing.
2. Install a knock-out drum to provide dryer motive steam.
3. Purchase an on-stream analyzer to monitor the seal drum off-gas composition.
4. Re-tube the surface condenser bundle with 12 gauge Titanium tubes.

I'm often tempted to slip away from reality with these sorts of suggestions. But in my heart and soul, I know what's really required—steely determination. If I make enough field measurements, spend countless hours talking to the operators on all the shifts, run tests at various vapor loads, steam conditions, and cooling water flows, examine all the operating data and devote hours and days analyzing the problem, eventually I'll come to a profound conclusion. Which is then typically proven wrong by a definitive plant test. And then I'll have to start all over again.

It's that very willingness to discard all my theories and begin anew, with an open mind that is the key to troubleshooting vacuum systems. I must never give up. Better death than defeat.

However, to be entirely honest, sometimes I have given up and recommended to my clients that they replace an outmoded steam ejector with a modern model. Then, I'll look back on the project years and decades later with a feeling of sadness and longing for a squandered opportunity.

"If only I had lowered the boot level in the pre-condenser, perhaps I could have unloaded the primary ejector," I'll think. But it will be too late. That was 20 years ago in a refinery that still exists only in my memories.

Sometimes, I'll start out my troubleshooting assignment to improve vacuum performance, surrounded by an enthusiastic group of young tech service engineers, older operators, and supervisors—all watching and waiting for the famous expert to solve their vacuum system problem. Three days later, I'm sitting all alone on the stairs leading to the 5th landing of the vacuum tower. It's starting to rain, and the evening chill is settling down across the refinery.

Now what? I'm almost out of ideas. I've tried almost everything I know without success. But there's always one more concept to explore. Did I remember to check the motive steam pressure to the third stage jet? Maybe it's a lot lower than it's supposed to be. It's always that one final measurement or observation that leads to the correct solution and on to that ultimate victory.

You just have to have faith that the correct solution is within your grasp if only you try long enough and with sufficient fortitude.

As an old Stillman, Leroy Wilkes, once said to me in a Texas City refinery in 1974, "Son, it's only a matter of time and determination."

Definition of Terms

Air Baffle – Means the same as vapor baffle.

Air Equivalent – The capacity of a jet is expressed in pounds of air. Other vapors, depending on their molecular weight, are converted to their air equivalent.

Back-Flushing – Flowing cooling water backwards through a condenser tube bundle to remove fouling deposits.

Barometric Leg – Drains a vacuum condenser.

Break and Pick-Up Mode of Jet Operation – Same concept as jet surging or hunting (see below).

Compression Ratio – A jet discharge pressure divided by its suction pressure.

Condensate Backup – Poor drainage causes increases in condenser outlet temperature.

Converging Section – First part of the diffuser.

Cracked Gas – The vapor vented from the seal drum.

Critical Discharge Pressure – The maximum pressure at the jet outlet, above which the sonic boost is lost.

Critical Mode of Operation – The jet is developing its sonic boost.

Diffusor – Section of jet where the process gas is compressed.

Diffusor Throat – Narrow portion of diffusor between the converging and the diverging sections.

Diverging Section – Last part of diffuser.

Dry Vacuum Tower – Ejector is on top of the vacuum tower.

Ejector – Correct term for steam jet.

Final Condenser – Operates at atmospheric pressure. Vents to the seal drum.

Fixed Tube Sheet Bundle – The unfortunate, but widespread practice, of condenser construction that precludes shell-side bundle cleaning.

Flame Arrestor – Prevents flame propagating backwards from heater into seal drum.

Hardness Deposits – Silicates and carbonates that restrict steam flow in ejector steam nozzles.

Helium Testing – Standard industry method to locate air leaks.

Hot Well – The boot at the bottom of the condenser.

Hunting – The sound an ejector makes when it is surging.

Inter-Stage Pressure – The pressure between the discharge of a jet and the inlet to the following condenser.

Iso-Entropic Expansion – Describes the conversion of heat to kinetic energy in a steam nozzle.

Jet – A device that produces a vacuum. Properly called an Ejector.

Jet Breaking – The jet has suddenly lost its sonic boost.

Leaking Turbine Shaft Seal – Source of vapor load to ejectors in surface condensers (i.e., air).

Mixing Chamber – The section of the jet where the process gas mixes with the motive steam.

Motive Steam – The working fluid that is used to compress the process gas. Any high velocity fluid can also be used (water, diesel oil, natural gas).

Nozzle Erosion – Wet, high velocity motive steam will enlarge the I.D. of a steam nozzle.

Nozzle Freeze-Up – Moisture in motive steam has frozen in the steam nozzle.

Off-Gas – Same as cracked gas.

Penetration – Main specification used for paving asphalt. Varied by controlling vacuum usually by throttling on motive steam to ejector.

Pre-condenser – Heat exchanger on top of the vacuum tower.

Seal Drum – Used to seal the bottom end of the barometric drain legs.

Seal Strips – Thin metal strips used to seal the air baffle to the interior of the condenser shell.

Sonic Boost – Mechanism to compress process gas in converging section of diffuser by reaching critical conditions (i.e., the speed of sound).

Sonic Velocity – Speed of sound corrected for flowing density of gas.

Steam Nozzle – Part of jet where the motive steam enthalpy and pressure is converted to speed.

Sub-Cooling Baffles – Used in surface condensers to increase NPSH available to boot pump (i.e, Net positive suction pressure).

Surface Condenser – A shell and tube exchanger with water on the tube side and mostly steam vapors on the shell-side. Has both a liquid and vapor outlet.

Surge – Periodic loss of the Sonic Boost. Also called hunting.

Thermal Cracking – Gas evolution due to time and temperature.

Throttling Mode of Jet Operation – A jet that is operating with motive steam pressure being controlled to control vacuum at less than maximum.

Tube Support Baffles – Used in condensers to support cooling water tubes.

Vapor Baffle – Part of condenser tube bundle that reduces vapor outlet temperature (same as the air baffle).

Velocity Boost – Mechanism to compress process gas in diverging section of diffusor.

Velocity Steam – Used in vacuum heater to suppress cracked gas evolution.

Vendor Performance Curve – All jet systems should conform to the jet's operating curve.

Venturi Effect – Conversion of pressure to acceleration.

Wet Steam – Moisture in motive steam. Degrades jet performance.

Wet Vacuum Tower – Pre-condenser located on the top of the tower, upstream of the first stage ejector.

Other Books by Author

- *Troubleshooting Refinery Processes*
- *Troubleshooting Natural Gas Processing*
- *A Working Guide to Process Equipment* (with Elizabeth Lieberman) – 3rd edition
- *Process Design for Reliable Operations* – 2nd edition
- *Troubleshooting Process Plant Control (Wiley Publications)*
- *Process Engineering for a Small Planet (Wiley Publications)*
- *Process Equipment Malfunctions*
- *Troubleshooting Process Operations* – 4th edition

The best method to purchase any of the above texts is from Amazon. *A Working Guide to Process Equipment* is the most popular of the above list. *Troubleshooting Process Operations* is the best text for refinery specific applications. Young engineers and operators find *Troubleshooting Process Plant Control* the most helpful. Check our website for details:

www.lieberman-eng.com

1

How Jets Work

No sane person is going to read this book unless they are troubleshooting vacuum system problems for steam turbine surface condensers or process vacuum towers. It's not a fun subject. As a matter of record, several of my closest colleagues have lapsed into insanity as a consequence of their interaction with ejectors, surface condensers, and seal drums.

I've never read a book, listened to a lecture, or seen a training video about vacuum systems. I have sometimes consulted vacuum system vendors, who have helped, but only up to a point. Still, my understanding as to how jets work is adequate for field troubleshooting.

1.1 The Converging-Diverging Ejector

A converging-diverging ejector is a two stage compressor, but with no moving parts. I've shown a sketch of such a jet

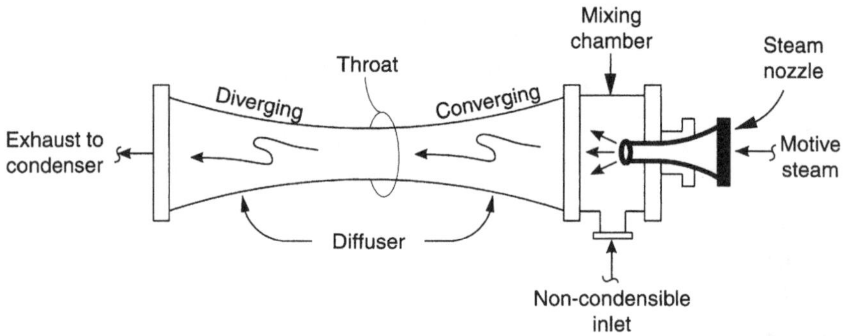

Figure 1.1 Components of a converging-diverging steam jet.

in Figure 1.1. If the jet has no moving parts, what is doing the work on the gas to compress it? The answer is the motive steam. And what property of the motive steam is doing the compression work? The answer is the velocity of the motive steam. This all has to be explained. It's rather complicated, but I'll make it simple for you.

You need to divide the ejector into three separate parts. I'll describe each part separately. Then, afterwards we can worry about their interaction.

- Part One – The Steam Nozzle – The steam nozzle is really small. You can probably hold a large one in your hand. It's much, much smaller in diameter than the steam supply line. It has a smooth, rounded opening. As steam expands through the steam nozzle, it accelerates from maybe 50 ft. per second in the supply line to perhaps 1,000 ft. per second at the discharge of the nozzle. The energy to accelerate the steam comes from two sources:
  1. Some from the pressure of the steam
  2. Most from the enthalpy (i.e., the heat of the steam)

I call the conversion of the steam pressure to kinetic energy, the Venturi Effect. This Venturi Effect is so efficient, that the pressure of the steam will drop as low as a few mm of Hg downstream of the nozzle in the mixing chamber (see Figure 1.1).

I call the conversion of the heat content of the steam to kinetic energy an Isoentropic Expansion. You can easily see what I mean. On your unit, check the temperature of the 150 psig (10 BAR) supply steam line. It's about 360°F (182°C). Now, check the temperature of the mixing chamber (which the nozzle exhausts into). It's about 90°F (assuming dry motive steam). What happened to the heat represented by the 270°F (360°F – 90°F) cooling of the steam? That heat was converted to speed. That's what Thermodynamics and the term, Isoentropic Expansion are all about:

- Thermo = Heat or Enthalpy
- Dynamics = Speed or Motion

- Part Two – The Converging Part of the Ejector – This is the half of the diffuser body that is downstream of the steam nozzle. It's perhaps 100 times larger than the steam nozzle. It's the front half of the diffuser shown in Figure 1.1. The motive steam enters the diffuser inlet at a velocity approaching sonic velocity or the speed of sound. The motive steam at this point already is combined with the off-gas from the vacuum tower or the upstream condenser vapor outlet. This off-gas has been drawn into the low pressure region of the mixing chamber created by the Venturi Effect of the motive steam.

  The narrowing cross-section of the converging section of the diffuser causes the motive steam (including the off-gas) to accelerate. The combined vapor stream reaches, and then exceeds, the speed of sound at or before the diffuser's narrowest section, called the throat. This is called critical flow or sonic velocity (see Figure 1.1).

As the flowing combined stream (i.e., steam plus gas) reaches sonic velocity, at or before the diffuser throat, it creates a pressure wave front which I'll call the Sonic Boost. This will compress the vapors by a factor of perhaps three or four to one. However, if sonic velocity is not reached at or before the diffuser throat, then the sonic boost does not develop and the vapors are not compressed at all.

- Part Three – The Diverging Portion of the Ejector – This is the half of the diffuser body downstream of the diffuser throat. It's the back half of the diffuser. The flowing vapors exit the throat into the gradually increasing cross-section of the diffuser. This causes the vapor to slow. The reduced kinetic energy of the vapor is converted into pressure. I call this the Velocity Boost. This will compress the vapor by a factor of about two or three to one. If the vapor velocity slows due to higher back pressure, this compression ratio is proportionally reduced.

  The combined effect of the sonic boost multiplied by the velocity boost is the overall ejector compression ratio. I have seen ratios, though rarely, as high as 12/1. A more common compression ratio is about 8/1.

  Loss of the Sonic Boost typically occurs for a wide variety of reasons, which is pretty much the subject of this book. When this happens, the ejector, which has been making a loud, roaring sound, will suddenly become much quieter. A sudden loss of vacuum will also result. The operators will then say, "The vacuum has broken." At the Delaware City Refinery, where I've been working recently (see Chapter 3), when the vacuum would break, the vacuum tower top pressure would jump from 4½ mm of Hg, to 10 or 12 mm Hg, and sometimes much, much more.

1.2 Interaction of Steam Nozzle with Converging-Diverging Diffuser

As the motive steam exhausts from the steam nozzle, it would be best for the steam to be moving at a maximum velocity. Since it's the velocity of the steam that is compressing the vacuum tower or surface condenser off-gas, and not its pressure or temperature, it's the velocity that always needs to be maximized. Malfunctions, such as erosion of the steam nozzle, hardness deposits in the nozzle, and low motive steam pressure will also reduce the nozzle exit velocity of the motive steam into the ejector's mixing chamber, shown in Figure 1.1.

As the combined off-gas plus motive steam flows into the converging section of the diffuser, we would wish the vapor to be moving at a maximum velocity, so that we would be able to reach sonic velocity at or before the diffuser throat. If not, the sonic boost will be lost. Back pressure from the diverging portion of the diffuser, along with low diffuser inlet velocity, will both reduce the vapor velocity in the diffuser throat.

As the vapor flows through the diverging portion of the diffuser, we would wish the vapor to encounter the least back pressure from the downstream condenser. If the exhaust flow from the diffuser does encounter excessive back pressure, then the back pressure will be transmitted back into the diffuser throat. This will not have very much effect on the overall compression ratio of the ejector, unless the velocity in the diffuser throat falls below sonic velocity. Then the sonic boost is totally and suddenly lost and the pressure in the surface condenser or vacuum tower will jump in a most alarming manner.

The factors that normally cause excessive diffuser exhaust pressure are warmer cooling water flow to the downstream condenser, fouling of the condenser, condensate back-up from the seal legs or condensate pump, loss of the sonic boost in the downstream ejector, condenser problems on the discharge of the downstream ejector, air leaks on the body of the diffuser, as well as many other possible problems.

And then, on top of all the above problems that cause a loss in sonic boost, are motive steam problems, such as excessive steam superheat, excessive motive steam pressure, too low motive steam pressure, air leaks, entrained liquids, excessive cracked gas, frictional losses in inter-connecting piping, and, again, the possibility for many other problems (see Chapter 3).

The reader can now understand why I have said that this book will not be fun to read. But, if you've got a vacuum system problem, my book is about this very topic. You're stuck with me. For better or worse, until success do we part.

1.3 Compression Ratio

When considering the performance of a vacuum jet, we must first consider the jet's overall compression ratio. To calculate a jet's compression ratio, use the following steps:

1. Measure the jet's suction pressure and convert to millimeters of mercury, as explained in Chapter 2.
2. Measure the jet's discharge pressure and convert to millimeters of mercury (mm Hg).
3. Divide the discharge by the suction pressure. This is the compression ratio.

It is not uncommon to find a proper jet developing an 8:1 ratio. More typically, jets will develop a 3:1 or 4:1 compression ratio. Any jet with less than a 2:1 compression ratio has some sort of really serious problem, but not necessarily with the jet itself.

1.4 Converging-Diverging Ejector

I'll remind the reader that this is a two-stage compressor with no moving parts. The first stage of the compressor is the converging section. The second stage is the diverging section. Each section develops a separate compression ratio. By compression ratio, I mean the outlet pressure divided by the inlet pressure. For example:

- Compression ratio of converging section = 40 mm Hg ÷ 10 mm Hg = 4.0
- Compression ratio of diverging section = 100 mm Hg ÷ 40 mm Hg = 2.5
- Overall compression ratio = (4.0) • (2.5) = 10.0

(Note – The 40 mm Hg cannot actually be measured on a process jet.)

These are typical design values for a properly performing jet operating within its design parameters of vapor loads, discharge pressure, and optimum motive steam conditions. Rarely do I observe in the field any single steam jet developing a ten to one compression ratio. I've conducted pressure surveys on a thousand steam jets in commercial service. Perhaps one or two percent develop a compression ratio of more than seven or eight to one. At a Conoco-Phillips Chemical Plant in Cedar Bayou, Texas, I've observed a single, small ejector, run at a 12 to one compression ratio. So anything is possible!

Now, I'm going to explain in a slightly different way how jets work. The first component of steam jets, is the steam inlet nozzle, as shown in Figure 1.1. High pressure motive steam flows through a specially shaped nozzle. It will help to think about your garden hose. Assume you have 40 psig city water pressure in your hose. As the water escapes through the nozzle, the 40 psig water pressure is converted to velocity. The greater the pressure of the water in the hose, the greater the velocity of the water escaping from the nozzle.

As the high velocity steam enters the inlet of the diffuser, shown in Figure 1.1, it starts to compress the non-condensable vapor drawn into the mixing chamber. I have read in some books, that the motive steam "entrains" the non-condensibles. This is wrong. The non-condensible gas flows into the mixing chamber for the same reason that any gas flows into the suction of any compressor. It flows towards the inlet of the diffuser because gas flows from an area of higher pressure to an area of lower pressure.

The kinetic energy required to accelerate the motive steam to sonic velocity as it enters the diffuser inlet comes from:

- The pressure of the steam
- The temperature of the steam
- The latent heat of the steam

This means that as the motive steam escapes from the steam nozzle, it cools and also partly condenses. Thus, it is normal to have water droplets blowing into the diffuser.

As you can see from Figure 1.1, the cross-sectional area of the diffuser diminishes as it approaches the diffuser throat. This forces the vapor velocity to increase. Upstream of the diffuser throat, sonic velocity is supposed to be achieved. If this happens, the jet is said to be in critical flow. The flowing vapor has exceeded the speed of sound. This creates a pressure wave front that I call the "Sonic Boost." It may compress the combination of flowing steam and non-condensables by a factor of four to one. To get the sonic boost, the velocity has to be above the speed of sound. As this velocity increases, the sonic boost compression ratio does not increase. However, if this velocity falls below the speed of sound, the sonic boost compression is instantly and totally lost. The converging section of the jet has stopped compressing the gas. The operators will say, "The Jet Has Broken," and observe a precipitous loss in vacuum. If you were standing next to the jet at this point, it would start to make a much quieter sound, which you are sure to notice. Then it likely will begin surging or hunting. More on surging later.

1.5 Velocity Boost

As the vapors pass into the diverging portion of the ejector, shown in Figure 1.1, the cross-sectional area of the diffuser increases. The vapor slows down. The reduction in kinetic energy is converted to pressure. I call this conversion of velocity to pressure, "The Velocity Boost." It may compress the combination of the flowing steam plus the non-condensables by a factor of two or three to one. The velocity boost is never entirely lost. It varies with steam pressure, and the vapor load and condenser back pressure. But it's always compressing the gas to some extent. The velocity boost is essentially the second, and smaller stage, of a two-stage compressor, with no moving parts. When the jet "breaks", the velocity boost continues working, even though the sonic boost has stopped completely.

Thermodynamics of Ejectors

Vacuum ejectors are two-stage compressors with no moving parts. The energy for both the first stage (sonic boost) and the second stage (velocity boost), comes from the kinetic energy of the motive steam. The faster the steam exhausts from the steam nozzle, the larger the compression ratio in the diffuser. The kinetic energy of the motive steam is derived by converting the enthalpy (both sensible heat and latent heat) plus the steam pressure to speed. While moisture in the motive steam upstream of the steam nozzle extracts heat, and thus kinetic energy from ejector, moisture downstream of the steam nozzle reflects an efficient conversion of heat to speed (an isoentropic expansion). That is, the conversion of enthalpy to velocity. The same principles apply to steam turbines, where the motive steam velocity spins the turbine wheel rather than the steam pressure.

1.6 Surging

Operators typically associate a sudden loss in vacuum (vacuum breaking) with the jets making a surging sound. When a jet is working properly, it makes a steady rather loud, roaring sound. If it loses its sonic boost, it will get quieter. But only for a moment. The loss of the sonic boost suddenly reduces the vapor load to the entire jet system. The jet discharge pressure is reduced because the downstream condenser is unloaded. This raises the velocity in the jet diffuser and the diffuser throat (see Figure 1.1). The lower throat pressure and higher velocity restores the sonic boost and the compression ratio. But this pulls forward the moles of gas that have backed up in the upstream vacuum system. The sudden increase in the gas flow increases the discharge pressure of the jet by increasing the gas flow to the downstream condenser. And then:

- The pressure in the diverging section goes up.
- The velocity (volume is inversely proportional to pressure) in the diffuser goes down.
- The velocity boost gets slightly smaller, which further raises the pressure in the throat.
- The velocity in the diffuser throat drops below the sonic velocity, and the sonic boost is therefore completely lost.
- Then the cycle repeats itself, and the operators will say, "Listen Norm! The jet is surging."

I used to think that only the first stage jets ever surged. But I now know that I was wrong. The bigger the jet, the louder the surging sound. If a downstream (2nd stage or 3rd stage) jet surges, then within one or two minutes, it causes the upstream jet to also surge. Why? Because the reduction in the downstream compression ratio increases the discharge pressure of the upstream jets. Once a jet's discharge pressure climbs above a certain value, it will lose its sonic boost, and hence begins to surge. The particular discharge pressure at which the sonic boost is usually lost, is called the jet's critical discharge pressure. I'll explain this later.

For now, note that once the large first stage jet surges, the smaller noise of the other jets surging is drowned-out. You'll only hear the 2nd and 3rd stage jets surging if you are present when the train of events is initiated, or at least, if you know exactly what you're listening for.

1.7 Critical Discharge Pressure

If the jet's critical discharge pressure is exceeded, then the jet will lose its sonic boost and surge. The converse is not true.

If the jet's discharge pressure is below its critical pressure, it may not develop its sonic boost.

How do we know the ejector's design critical discharge pressure? It's listed on the jet's data sheet or performance curve. It's the same as the maximum jet discharge pressure. Above this pressure, the jet can't be expected to operate on its vendor supplied performance curve. No matter how small the load, no matter how dry the steam, no matter how good the jet's mechanical condition, exceeding the jet's design critical discharge pressure is usually going to cause the jet to surge and eventually lose its sonic boost.

The most common cause of high jet discharge pressure, other than the downstream jets surging, are condenser related problems, which I've described in subsequent chapters 9, 11, 12, 13, and 15.

1.8 Observing the Conversion of Heat to Velocity

The conversion of enthalpy to kinetic energy is not confined to steam ejectors or even other sorts of process equipment. We can see this happen every day in the world around us.

My favorite example is watching air flow across the top of the wing of an airplane. But how can I see air? Well, as the air flows across the top of the wing, due to the shape of the wing and the angle of attack, the air accelerates. The energy to accelerate the air does not come from the plane, but from the barometric pressure and heat content of the air itself. Thus, the air cools as its kinetic energy increases. Depending on the relative humidity of the air prior to this cooling effect, the air may drop below its dew-point temperature. Moisture, in the form of a fog (a colloidal suspension) will make the flowing air quite visible.

The first time I set foot in a refinery was in 1965, in Whiting, Indiana. I recall encountering a large 200 psig steam leak. Now, 200 psig is about 380°F. Yet the blowing steam I accidentally walked through, while warm, was not 380°F. It seemed like the steam was cooling as it escaped through the leaking pipe. And so it was. To some extent, the heat content of the steam had been converted to velocity.

Here's an experiment you can perform at your own convenience, in the privacy of your own home. Depress the valve on the stem of a car tire. Doesn't that air feel somewhat cooler than the tire itself? (I've measured this at about 20°F temperature reduction.)

I have another way of visualizing the conversion of heat to kinetic energy. That is, the energy to compress vapor, whether derived from a steam ejector, or from a steam turbine, is extracted from the velocity of the steam. But when we raise steam, the energy consumed goes largely into latent heat and increasing the sensible heat content of the boiler feed water, rather than the pressure. I've discussed this in great detail in my book, *Process Equipment Malfunctions*, McGraw Hill, 2011.

The world that surrounds us is a lot more complex than how we first perceive it as children. I'm 70, yet I'm still learning new things. Not about how distillation trays malfunction, or surge in centrifugal compressors. What I have in mind is, how do kites fly; how can a sailboat tack into the wind; what causes ice to melt when it's colder than 0°C; why diesel engines don't need spark plugs but gasoline engines do; how does an electric motor work; what causes water hammer in my attic piping; what draws air into my fireplace; and, how does my aquarium air lift pump circulate water? These familiar devices turn out to operate on basic, but not necessarily, obvious principles.

Just like steam ejectors.

1.9 Jet Discharge Pressure

The jet suction pressure is a function of the following factors:

- The overall jet compression ratio.
- The jet discharge pressure.

The jet discharge pressure is controlled by the downstream condenser pressure. The minimum condenser pressure corresponds to the condensing pressure of the steam at the condenser's vapor outlet temperature. For example, let's say that the condensing pressure of pure steam at 120°F is 87 mm Hg. If the condenser vapor outlet temperature is 120°F, then the lowest pressure we could expect to measure at the condenser vapor outlet would be 87 mm Hg.

Let's further assume that the pressure drop from the upstream jet discharge through the condenser discharge is 10 mm Hg. The jet discharge pressure would be 97 mm Hg. Let's also say that the sonic boost is equal to 3.60. The velocity boost is assumed to be equal to 2.5. The overall compression ratio is then:

- $3.60 \times 2.5 = 9.0$

The jet's suction pressure is then:

- 97 mm Hg $\div$ 9.0 = 10.8 mm Hg

1.10 Reducing Primary-Jet Discharge Pressure

Let's say that a jet is already in its critical-flow mode. It is already benefitting from both the sonic boost and the velocity boost. What, then, will be the effect of a reduction in the jet's discharge pressure on the jet's suction pressure? Answer: Not very much. If a reduction in discharge pressure is made on a jet that is *not* working in its critical mode, there will always be some benefit.

But if the jet is already in critical flow, reducing the pressure downstream of the diffuser throat cannot significantly raise the flow of gas into the diffuser throat. I know. I've tried. Twice I have added a third-stage jet to an existing two-stage jet system. The discharge pressure from the second stage jet dropped by 500 mm Hg. The discharge pressure of the first-stage jet dropped by 160 mm Hg. The suction pressure to the first-stage jet dropped by perhaps 2 mm Hg.

The critical discharge pressure for each jet is determined experimentally by the manufacturer. It is usually noted on the jet specification sheet (see Figure 3.2). My experience indicates that exceeding this critical jet discharge pressure by the smallest amount will force the jet out of critical flow. Or, the way I've experienced it, will cause the jet to suddenly surge a few times, and then lose its sonic boost.

The inverse is not always true. A jet with less than its critical discharge pressure may not pick up its sonic boost. Operationally, this business of gaining or losing sonic boost is a rather dramatic and often terrifying effect.

1.11 Bypassing First Stage Ejectors

I've encountered the same problem twice, at the Sinclear Refinery in Wyoming, and also at a small asphalt plant in Corpus Christi, Texas. In both cases, especially in Corpus, shutting off the motive steam to the first stage ejector improved the vacuum tower pressure.

The problem arises either due to excessive back pressure from a malfunctioning downstream condenser (in Wyoming) or mechanical deficiencies with the jet itself (Corpus Christi). In either case, the jet was only boosting the gas pressure by five or ten mm Hg.

Now, one would think that every little bit of compression is helpful, so run the malfunctioning jet anyway. But this is far from true. I recall that at the Corpus Christi asphalt plant, that when I shut down the motive steam to the first stage ejector, the vacuum tower top pressure dropped by a few mm Hg. The reason was that I had greatly reduced the heat duty to the first stage condenser. The condenser vapor outlet temperature dropped, as did the condenser pressure. The tower pressure was now floating on the first stage condenser pressure. I measured the delta P across the idled ejector, which was about three or four mm of Hg. I then opened the bypass around the ejector, and the vacuum tower pressure dropped by a few more mm of Hg. This was really good.

Not only had I substantially reduced the vacuum tower top pressure, but I had also saved several thousand pounds of wasted motive steam used on the first stage jet.

I did the exact same thing at the Sinclear plant that year with initially the same wonderful results. But then, I tried to run an experiment. I wondered how, without the first stage jet in service, the ejector system would perform at a much greater vapor load. So I opened a ¾″ bleeder at the inlet to the first stage condenser to suck in air. Not for just a few seconds, but for several minutes. Suddenly, the second stage ejectors (which were now the lead ejectors) started to surge, and I lost a tremendous amount of vacuum.

"Okay," I thought, "This is bad. I've made a mistake." So, I closed the ¾″ bleeder and nothing happened. The jets did not stop surging, and the vacuum was not restored. I then put the first stage ejector back into service, which did help, but the vacuum was now worse than before I had done anything.

I've thought and thought about this incident, and have concluded that the answer is that there is no answer. I just should have quit while I was ahead. I must have oxidized iron sulfides or hydrocarbon deposits inside the first stage condenser and that somehow degraded its performance. I have the same problem when I play roulette.

2

Making Field Measurements

The ability to carefully make field measurements with accuracy and to correctly interpret their meaning is the ultimate key to troubleshooting vacuum system malfunctions. I always insist on making all field measurements myself. Like John Henry who died with a hammer in his hand, I'll go out holding my digital vacuum pressure gauge.

2.1 Getting Started

Before you begin troubleshooting your vacuum system, you will need three items:

1. A digital vacuum manometer. I suggest you purchase three as they are not particularly robust. I use a Meriam Instrument Model A1500MH, S/N 181870-LI employing a nine volt battery. The cost is about $500 each. (Range 0 – 1,500 mm of Hg.)

2. An infrared surface temperature gun, available at most hardware stores for \$30 or \$40. (Range 0°F – 380°F.)
3. Steely determination and unlimited time.

The manometer is affected by radio signals and other electric interference. It tends to read low by a few mm of Hg as the vacuum gets very good. For example, at a reading of 4 mm Hg, the real vacuum may be 6 mm Hg. I'll discuss later in this text how to make very accurate vacuum measurements at low vacuums. Note that no correction for changes in barometric pressure is required with this sort of digital manometer.

In addition to making temperature and process measurements, you will want to sample the off-gas from the seal drum. This sample should be taken in a container without metal parts. Any such metal parts will catalyze the oxygen plus H_2S conversion to sulfur. As explained later in this text, you do not want to lose the oxygen content of your sample. It's most likely best to measure O_2 locally in a Drager tube.

I always make all my measurements myself. It's not that I don't trust the operators. I don't trust anyone. Mainly, though, troubleshooting vacuum systems is so complex that in practice, I'll accidentally stumble across the solution in the course of making my field measurements.

If you lack an intermediate pressure point because there is only a plug and no fitted valves, just pull the plug and install a valve using the following procedure:

- Don't ask permission. Just do it.
- Don't spend more than two seconds pulling the plug and placing the screwed nipple and valve in place.
- Don't drop the valve.

The seal drum off-gas is far too fuel rich to ignite if you do the job quickly.

Find the performance curves for your jet system. Often, you will have to call the equipment vendor for this data. Also, locate the condenser data sheets and the critical discharge pressure for the ejectors and their design motive steam pressure (the latter will be on the ejector name plate).

Pretty much, most of the temperatures you will be getting with the infrared thermometer will be in the 80°F to 160°F range. Therefore, the skin and actual process temperatures will be pretty much the same. Note that the infrared thermometer cannot be used on shiny, reflective, stainless piping. Always look for the highest reading you can obtain at any particular location.

If you find a small localized cold spot – and I mean really cold, like 40°F – you have likely identified an air leak. Air sucked into a vacuum system auto-refrigerates.

Listen carefully for the ejectors making a surging, roaring, or hunting sound. That's bad. The jets are losing their "sonic boost" (explained in Chapters 1 and 3).

The louder the jet sounds, the better the jet is working. A quiet jet may be doing very little compression work.

The main concept though is not to be discouraged. Here is my mantra:

- No one can help me.
- It's a matter of life and death.
- Man is the greatest factor in the universe; he can do anything.

It's a good idea to talk to the plant operators and learn from their experiences and views. Don't believe everything they tell you; but don't entirely discount their observations either. Don't be afraid of trying out operational changes. The price we pay for success is the willingness to risk failure.

I know it's a pain, but read this whole book. Your answer is probably somewhere in the following pages. Good luck and happy hunting.

Getting started on any assignment is, for me, always the most difficult part of the job. For vacuum systems, I proceed differently, depending on whether or not the system has a pre-condenser. For condensing steam turbines, the surface condenser is much the same as the vacuum tower pre-condenser.

Let's assume I have a pre-condenser or a steam turbine surface condenser. I will then make two measurements first:

- The condenser vapor (not the liquid) outlet temperature.
- The pressure at the condenser vapor outlet nozzle.

I'll then look up the vapor pressure of water on my steam table (see Table 2.1). If my measured pressure is near or a bit below the vapor pressure of water at the condenser outlet temperature, then I'll draw the following conclusions:

Table 2.1 Vapor Pressure of Water in PSIA.

Temperature, °F	Vapor Pressure Water, PSIA
80°F	0.5
90°F	0.7
100°F	1.0
110°F	1.3
120°F	1.7
130°F	2.2
140°F	2.9
150°F	3.7
160°F	4.7

- The vacuum I observe is only a function of the condenser efficiency.
- Improvements in the performance of the downstream ejectors or condensers cannot significantly improve vacuum.

Incidentally I have assumed, and this is most always the case, that a vacuum tower with a pre-condenser uses steam in its feed heater coils or as bottom's stripping steam.

For temperature measurements, I'll use my $40 infrared surface thermometer purchased at the local hardware store. For pressure, I'll use a $500 digital vacuum gauge purchased from a lab supply catalogue. If I don't have a local pressure connection, I'll quickly remove the plug and install a ½ inch gate valve.

If there is no pre-condenser, then I'll measure the pressure before and after each jet. Also, I'll measure the skin temperature at the vapor and liquid outlet of each condenser. This can take an hour or two. Also, if the vacuum system is not limited by the vapor pressure of water at the pre-condenser vapor outlet temperature, then I'll also have to make these additional measurements as defined below.

If the system vacuum is limited by the water vapor pressure at the pre-condenser outlet, then I will check the condensate drain leg temperature. If this temperature is much colder (10°F or more) than the vapor outlet temperature, then the condenser is suffering from either:

- A leaking vapor or air baffle (Chapter 15).
- Condensate back-up (Chapter 12).

These malfunctions are discussed in great detail latter in this text.

Alternately, both the vapor and liquid outlets (condensate drain) temperature may be excessive. Then the problem may be:

- Lack of cooling water flow (Chapter 11).
- Exchanger fouling (Chapter 9).
- Overloading the condenser with excessive steam and/or condensable hydrocarbons (Chapter 6).

I've given examples of all of these malfunctions in this text.

If the vacuum is not limited by the vapor pressure of water at the pre-condenser or surface condenser vapor outlet temperature, then the scope of my work is vastly expanded. In addition to the temperature and pressure surveys I've just described, you will want to have:

- The "critical" (defined in this text) discharge pressure of each ejector.
- The ejector performance curve (provided by the equipment vendor).
- A measurement of three flows:
  a. Seal drum off-gas (Chapter 6).
  b. Seal drum hydrocarbon condensibles.
  c. Steam used in the turbine or vacuum tower.
- An analysis of the gas and hydrocarbon liquid produced in the seal drum.
- The design (or actual, if metered) motive steam consumed in each ejector.
- The actual motive steam pressure and temperature to each jet is measured just upstream of the jet (i.e., NOT the steam supply pressure).

A large portion of this book is devoted to explaining how to measure these parameters. Half the time, in working to obtain all of this data, I will stumble across the cause of the bad vacuum. That's why I've saved my most important recommendation for last. You have to make all the measurements and observations I describe in my book yourself:

"You've got to walk that lonesome valley."
"You've got to walk it by yourself."
"No one else can walk it for you."
"You've got to walk it for yourself."

(Song from "Ox Bow Incident," starring Henry Fonda)

2.2 How to Unscrew Steel Plugs

Our work may involve making pressure measurements on equipment that no one has ever run a pressure drop survey on before. Many ¾" or one inch steel screwed plugs are not easily removed. Here's what you'll need to do the job safely:

- Two, 12" pipe wrenches.
- Two, two to three foot long pipes.
- Sledge hammer.
- A can of WD-40.
- Someone to help you.

First spray the recalcitrant plug with WD-40. Then, hit it a few sharp blows with the sledge hammer. I don't think this does all that much good, but it makes me feel better.

I'll then try to turn the plug (counter-clockwise) with my wrench. If this doesn't work, I'll bang really hard on the wrench handle with my sledge hammer. If I still can't unscrew the plug, I'll ask for help.

The stronger guy (or gal) will take the shorter pipe and slip it over the end of the 12" pipe wrench as an extension handle. The other participant will place his wrench and pipe extension on the valve fitting to keep it from breaking off (or unscrewing). If this is unsuccessful, get longer pipe extensions.

I was once working for Total in France with Oliver, a PhD Chemical Engineer. We were confronted with a 1" plug that the plant operators could not unscrew. I used a twelve foot scaffold pole as an extension, and with one hand removed the stubborn plug. Oliver thought this was the most wonderful technical achievement he had ever witnessed.

2.3 Effect of Barometric Pressure on Indicated Vacuum

As barometric pressure or atmospheric pressure increases, there ought to be a very small reduction in the upstream vacuum—meaning the upstream pressure of the vacuum system should get worse. Typically, this effect is far too small to observe. However, on certain vacuum systems, the indicated vacuum appears to greatly improve as the ambient pressure increases.

Ambient pressure (or barometric pressure) normally increases about ten mm of Hg as it gets cold outside in the evening in Big Springs, Texas. And the indicated vacuum tower pressure declines (i.e., it improves).

Actually, the vacuum tower top pressure has not changed at all. It's just that the vacuum pressure indicator is not compensated for the ambient pressure.

My portable electronic digital vacuum gauge, which I use for all my field pressure surveys, is not affected by the external vacuum tower pressure. The vast majority of panel pressure indicators are compensated for, or unaffected by ambient pressure. However, on some of the older installations, I still see this problem.

It's very confusing because very often, ambient conditions do affect the vacuum by changing the cooling water or the air cooler temperatures.

At the Big Springs Refinery in West Texas, I placed my portable, digital, vacuum pressure indicator on the top of their vacuum tower, adjacent to the top pressure transmitter. As the ambient temperature dropped in the desert, and as a result, the barometric pressure increased, the real vacuum as shown on my portable gauge did not change, even as the indicated vacuum on the panel became better.

I saw the same problem last year on an older installation at the Hovensa Refinery in St. Croix. It's not too likely you or I will see this sort of malfunction very often. But just in case, you should keep this problem in mind.

2.4 Use of Piccolo

The very high velocities that the ejector is subject to expose the steam nozzle and diffuser throat to excessive wear (Figure 1.1). Low quality steam will accelerate this erosion. A gradual loss in vacuum may be due to enlargement of the ejector clearances. It is a good practice to caliper these clearances when the system is out of service. A fuller description of ejector operations, steam requirements, and pressure capabilities can be found in Robert Frumerman's article "Steam Jet Ejectors," appearing in the June 1956 issue of *Chemical Engineering*.

To determine if a jet has suffered mechanical deterioration, proceed as follows:

1. Take the jet off-line and vent it to the atmosphere. Insert a blanking plate in the mixing chamber inlet.
2. Install a vacuum pressure gauge. Measure the vacuum developed.
3. Compare the observed vacuum with the predicted vacuum at zero air flow as shown on the manufacturer's jet performance curve.
4. If the vacuum is below the predicted curve value, the jet is mechanically deficient. However, if this is not the case, continue to the next step.

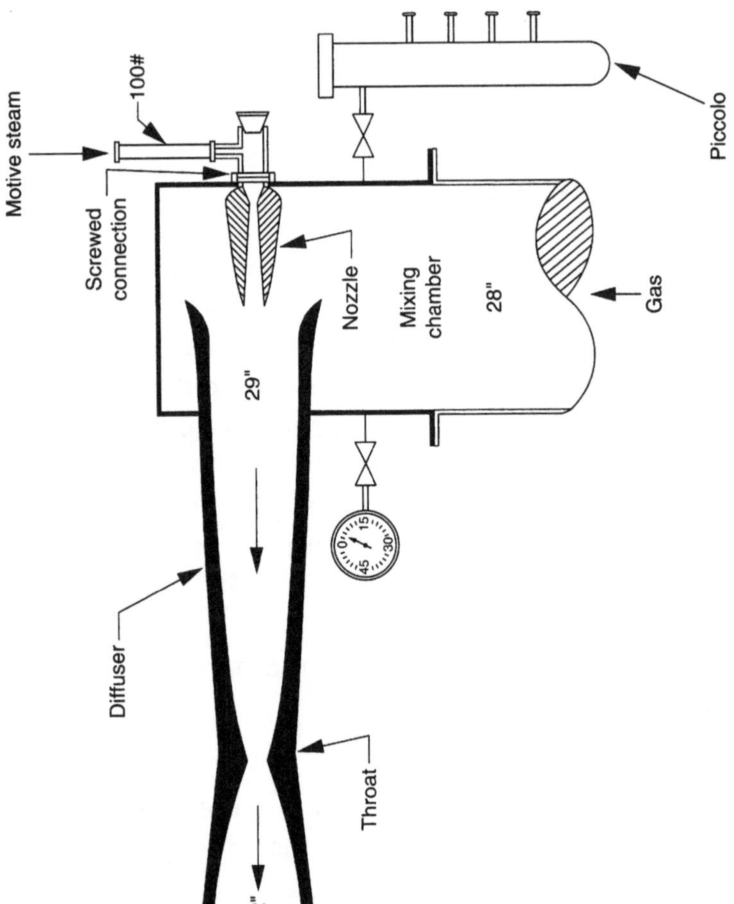

Figure 2.1 A vacuum jet or ejector equiped with a piccolo used for off-line testing.

5. Install a "piccolo", as shown in Figure 2.1. The piccolo is nothing more than a pipe with calibrated holes. It is also obtained from the jet manufacturer. Remove one plug at a time. From a chart that comes with the piccolo, read the air flow to the jet based on the pressure in the mixing chamber and the hole size. Using the piccolo, you can generate a jet performance curve and compare it to the original performance curve.

Very often, it is erosion of the jet's nozzle that accounts for loss of vacuum due to deterioration of the jet. To prove this, simply throttle the motive steam inlet (with the jet vented to atmosphere) and the vacuum will improve. Sometimes, steam leaks around the steam nozzle into the mixing chamber. Sealing these leaks will also enhance vacuum. Hardness deposits forming in the steam nozzles due to poor quality motive steam are also an extremely common problem.

2.5 Measuring Deep Vacuums using an Hg Manometer

For any vacuums better than 120 mm Hg (or 25 inches of Hg, at sea level), an ordinary vacuum pressure gauge will not be accurate enough for technical purposes. An absolute mercury manometer, as shown in Figure 2.2, is needed. All that is required to make this simple device is a length of glass tubing bent into a U-tube shape. One end is sealed and the other end left open. Dry, clean mercury is then poured into the open end. The closed end of the U-tube is easily evacuated of air by tipping the glass U-tube on its side. A little jiggling will work out the last air bubble. The overall length of the tube will be about 8 inches. The mercury should wind up about 1 or 2 inches high in the open end of the U-tube.

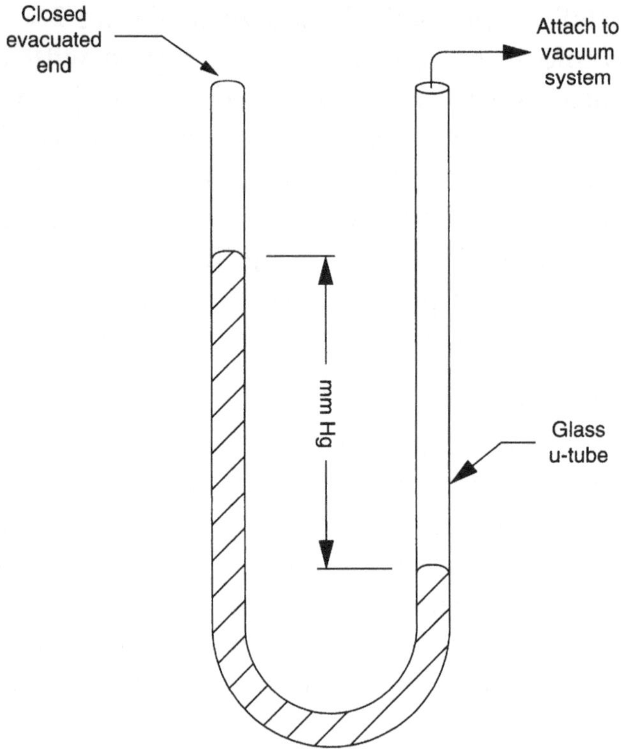

Figure 2.2 An absolute vacuum mercury manometer.

To read the vacuum, the mercury level at the closed end must be pulled down by the vacuum a bit below the top closed of the tube. The difference in mercury levels between the closed and open ends of the U-tube is the precise mm Hg vacuum. Wet mercury on the closed end of the "U" tube will give completely wrong results due to the vapor pressure of water.

2.6 Measurement of a Deep Vacuum without Mercury

Above 30 or 40 mm Hg of vacuum, an ordinary digital battery powered vacuum gauge is sufficient. However, only a vacuum mercury manometer is accurate in the 5 to 15 mm Hg range. As long as the mercury is dry, such a gauge gives excellent results. But few of my clients now permit the use of mercury in the field due to the toxicity of mercury.

I have developed an alternate method that works almost as well as the vacuum mercury manometer. It's really the same idea. See Figure 2.3. Fill the closed end of the tubing with baby oil or mineral oil. Make sure you get the last bubble of air out of the closed end of the tubing. Always use fresh oil to prevent moisture contamination of the oil. Fill the entire glass U tube with oil. Connect the open end of the glass tubing to the vacuum pressure point being measured and open the connecting valve very slowly.

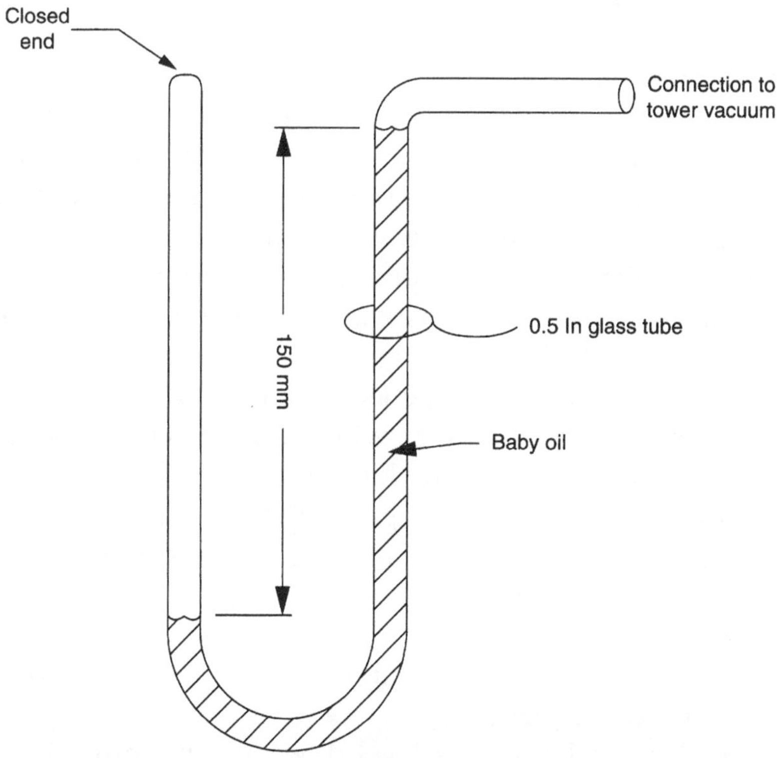

Figure 2.3 Oil filled deep vacuum manometer.

The specific gravity of baby oil is about 0.9 s.g. The specific gravity of mercury is 13.6. Mercury is 15 times denser than baby oil. Referring to Figure 2.3, divide the 150 mm of elevation difference in the manometer legs by 15 to determine that the vacuum is 10 mm Hg. For vacuums above 40 mm Hg, the oil-filled mercury manometer is not practical, as it would be about 30 inches long, which is an awkward length for field work.

Do not use fluids that are more volatile than baby oil, as they will create a vapor pressure at the closed end of the glass U tube. This will result in a measurement of a better vacuum than really exists. Water is also too volatile at moderate ambient conditions for this purpose.

2.7 Measuring Condensibles in Feed to First Stage Ejector

If your pre-condenser is upstream of the first stage jet, then the load to the first stage jet is pretty close to the evolved wet gas from the seal drum. Certainly, there's some additional condensation between the discharge of the first stage jet and the seal drum. However, there is no way to estimate the condensate so formed, as distinct from the pre-condenser liquid effluent, and it will be relatively small, so I'll just ignore this contribution to the first stage jet's load (see Figure 3.1 in next chapter).

However, if the first stage jet is taking flow directly from the top of the vacuum tower, with no intervening condenser, then the condensable load to the first stage ejector may be very large as compared to the wet gas. But, how does one measure this condensable load? Actually, here's a pretty simple and reasonably accurate way:

- Step 1 – Find the liquid level in the seal drum of the gauge glass.
- Step 2 – Draw this level down to minimum. I'll just bypass the LRC in the field to do this step.

- Step 3 – Shut down the oil-side pump on the seal drum (but not the water condensate pump).
- Step 4 – Watch the level creep up in the seal drum oil-side gauge glass. Let it increase from its minimum level to its maximum level. That is, about one to three feet for most seal drums. Record the elapsed time interval.
- Step 5 – Using the seal drum vessel sketch as a guide, calculate the volume of hydrocarbon condensate that accumulates in step four. (Don't forget about the internal baffles in the seal drum.)
- Step 6 – Turn the pump back on and obtain a sample of the hydrocarbon condensate.

Check the gravity of the condensate. Typically, the s.g. will be 0.65 to 0.70. Send the sample to the lab for analysis (D-86 boiling range distillation). Material boiling below 350°F is naphtha. You may assume a molecular weight for the naphtha of 100.

Material boiling between 350°F to 500°F is jet fuel or light diesel. You may assume a molecular weight for this kerosene type of material of about 150.

Material boiling above 500°F is likely to represent entrained LVGO (i.e., liquid) and is not actually contributing to the primary ejector vapor load.

My experience is that most of the seal drum hydrocarbon liquid collected is light jet fuel or kerosene. Most of my clients send this stream to slop. Routing it back to the crude desalter, along with the vacuum seal drum water is a more intelligent option.

2.8 Identifying Loss of Sonic Boost by Sound

A jet that is developing its sonic boost is said to be in its critical mode of operation. To be in a critical mode of operation, the jet's discharge pressure has to be below its critical discharge pressure. Once, I was working at the Murphy Oil Refinery near my home in New Orleans, Louisiana. Both the first and second stage jets had excess capacity. But the third stage jet was bogging down due to gas overload. It started to surge as it periodically lost and regained its critical mode of operation.

My problem is I don't hear very well. This is the result of forty-plus years in process plants without wearing protection (ear plugs). The smaller the jet, the less audible its surging sound is going to be.

As the tiny third stage jet surged, it caused the discharge pressure from the medium sized second stage jet to exceed its critical discharge pressure. The second stage jet now also began to surge, which I did not hear, either. But Kelly, the younger outside operator who was helping me, did hear. The only thing I could hear for myself was the roaring surge of the giant first stage jet, which was surging due to exceeding its critical discharge pressure, because the second stage jet had previously lost its sonic boost.

"Mr. Norm," young Kelly said, "The second stage jet. Listen. It's breaking. It's making that there hunting sound again."

"Right you are, Kelly. I'll order a new and larger replacement," I said.

And so I did, which did no good at all. This, of course, did not make my client too happy with me.

Only later, did Gerry, an even younger and keener eared process engineer, hear that it was the small third stage jet that had surged prior to the second stage jet. I could have heard it myself if I had stood right next to the ejector and listened really carefully.

2.9 Identifying Air Leaks

I have developed a number of ways to determine the location of air leaks. If sample results indicate over 20% N_2 in the seal drum off-gas, then a major air leak is indicated.

The first step is to consider the ratio of O_2/CO_2. If this ratio is small, then the air leak is most likely in the vacuum heater transfer line. However, if this ratio is large, the air leak is in the overhead ejector system. Assuming such is the case, the location of such air leaks can be determined by several methods. The simplest method is to spray aerosol shaving cream around the suspect leak and see if the cream is drawn into the pipe or flange. If the condenser or ejector can be isolated, then pressure up the equipment with steam and look for steam blowing out of the leak.

In my experience, most such vacuum leaks are associated with poor workmanship in making up the flanges. Usually, that's due to careless installation of gaskets. With this in mind, you should proceed as follows:

- Wrap each flange with duct tape (which is always a good operating practice, anyway).
- Punch a small hole in the duct tape.
- Spray aerosol shaving cream around the hole.

A more comprehensive method to find such air leaks is to hire a company to spray helium around areas of potential leaks. The helium will be drawn into the leak and appear a few moments later with the vent gas from the seal drum. A continuous Helium detector at this location will allow the general area of the air leak to be identified. I've done the same thing by myself, but using air conditioning Freon and a $600 Freon leak detector kit, purchased from my local air conditioning supply store.

Finally, as I've previously suggested, just look for cold spots around fittings, piping, and flanges. When air is drawn through such a leak into a vacuum system, it will cool by perhaps 50°F. But how does one "see" a cold spot on a screwed fitting? Simple. Just look for moisture dripping off such a flange or fitting. This won't work in areas of low humidity, but it works quite nicely in New Orleans and Houston.

I've read in other textbooks that there are methods to find air leaks in vacuum systems by some sort of "ultrasound detection device". I've never seen one used myself and have never met anyone else who has used one. Background noise in a refinery would seem to preclude the practical use of such a method.

2.10 Air Leaks in Flanges

This is an especially important subject to me, as I've had this problem on too many occasions. Such flange leaks should never happen. It's all down to sloppy pipefitting. The common causes are:

- The pipefitters have not pulled up the flanges evenly. They have over-tightened the bolts on one side of the flange first. Bolts must be tightened in a well defined sequence to avoid such failures.
- The gasket has never been installed. This happened at the Amoco Sugar Creek Refinery. An internal flange inside the seal drum leaked because no gasket was used. Being inside the seal drum, the leaking flange was not found during the steam pressure test. The pipefitters had thought that since the flange was inside a vessel, a small leak was not a problem, and no gasket was needed.
- Pieces of the old gasket are left sticking to the flange surfaces when a new gasket is installed. This is the most common (and most inexcusable) cause of flange leaks. I'm still angry about such an incident that occurred in 1974 at Amoco's Texas City refinery on my alky unit depropanizer.
- A paper type gasket has been used. Such gaskets are only intended for short-term use during turnarounds, when using slip blinds (or spades in Europe). They are never intended for continuous use on operating units. I learned all about this in a refinery in Lithuania in 2006.

2.11 Vacuum Measurement Units

I was once captured by aliens and transported to their home planet, which rotates around the star, Sirus.

"Earthling," asked their many-eyed leader, "Explain how sub-atmospheric pressures are quantified on the 3rd planet from your home star."

"Well, it's kinda complicated," I replied.

"Yes! Yes! We know. That is why we have brought you here. So we may gain insight and achieve understanding. Speak now Earthling, or die."

"Oka. It's like this. In America, we use the system of inches of Hg vacuum,

This is how it works:

- At sea level, atmospheric pressure is zero inches of Hg. And full vacuum is 30 inches of Hg.
- At a higher elevation, like Denver, atmospheric pressure is zero inches of Hg. And full vacuum is 28 inches of Hg.
- Thus, a vacuum of 26 inches of Hg at my home in New Orleans (at sea level), is the same pressure as 24 inches of Hg in Denver. Both are four inches of Hg above a pressure of absolute zero."

The multi-eyed green aliens glared at me with disbelief and hostility.

"Okay. Don't get angry," I said. "We've got a simpler system of measurement too. It's based on mm of Hg. There are 25.4 mm per inch. Here is the alternate method:

- In Denver, zero mm of Hg is full vacuum. And atmospheric pressure is 709 mm of Hg.
- Thus a vacuum either in Denver or New Orleans of 102 mm of Hg is the same pressure.

- And this 102 mm of Hg is the same pressure as 26 inches of Hg vacuum at sea level in New Orleans and 24 inches of Hg in Denver."

"Creature from the blue planet, we think the 'mm of Hg,' is best," the hostile allen said.

"I agree. But we have other systems as well. For example:

- Minus 13 psig at sea level is the same as 26 inches of Hg or 102 mm of Hg.
- Minus 12 psig in Denver is the same as 24 inches of Hg, or 102 mm of Hg."

My captors turned from green to brown, just like the lizards in my backyard. What did this signify?

"And then we have another system," I continued to explain. "Minus 0.5 BARG at sea level is equal to 380 mm of Hg, or 15 inches of Hg vacuum, or minus 7½ psig.

'At a higher elevation, in Denver, 0.5 BARG is equal…"

"Stop, Earthling!" screamed my captors.

"And then there are inches of water vacuum used at smaller vacuums. And then there are inches of water draft used in fired heaters. Also, shall I explain negative kpa and millibars of vacuum?" I continued.

"No earth creature," intoned the many-eyed commander. "Go home in peace. We have decided not to invade your blue planet. It's too complicated."

For me, I avoid all these complications. I make all my measurements with my digital pressure gauge, which reads in mm of Hg and needs no correction for elevation. I also carry out all calculations in mm of Hg, and would advise the reader to do the same.

"Earthman," they said, "The environment of this planet is so degraded, it will not be habitable by intelligent life forms for a minimum of another 128 solar cycles. It's not worth our trouble to invade."

And then they were gone, without actually inquiring about our vacuum systems. However, you may find Table 2.2 and Table 2.3 helpful.

Incidentally, I made up the part about my being captured by aliens and then transported to their home planet. The truth is that once they realized how complicated all of this stuff was, they released me almost immediately.

Table 2.2 Vacuum Measurement Systems.

Actual atmospheric pressure = 29.97 in Hg		**Actual atmospheric pressure = 25.00 in Hg**	
in Hg	mm Hg	in Hg	mm Hg
0	760	0	635
5.0	633	5.0	508
10.0	506	10.0	381
15.0	379	15.0	254
20.0	252	20.0	126
25.0	125	25.0	0
29.97	0		

Table 2.3 Methods of Expressing Vacuum.

Very Deep Vacuum – Usually expressed in mm Hg. There are 760 mm Hg at atmospheric pressure. Zero mm Hg is absolute zero pressure.
Moderate Vacuum – Expressed in inches of Hg. There are 29.9 inches of Hg at absolute zero pressure, and zero inches of Hg at atmospheric pressure. A vacuum of 29 inches of Hg therefore equals a vacuum of 25 mm Hg.
Minor Vacuum – Expressed in inches of water. Often called draft. Thirty inches of water draft corresponds to a vacuum of about two inches of Hg.
Absolute Pressure – To avoid confusion, the designer should not use psia in designating equipment requirements. A pressure of 1 psia corresponds to a vacuum of 50 mm Hg or 28 inches of Hg.

3

Tabulation of Vacuum System Malfunctions

This book is mainly about the most difficult piece of plant equipment to troubleshoot that process operators or engineers must deal with: the **STEAM EJECTOR**. It's made up of three parts: a driver and a two stage gas compressor. I'll summarize the main components of the ejector that I have discussed in greater detail in the first chapter.

- The Driver – That's the steam nozzle. The heat and the pressure of the steam is converted to velocity. An area of low pressure and super-high velocity is created. The low pressure steam sucks in the off-gas.
- The Converging Section – The steam and off-gas rush into a piece of pipe, which gradually gets narrower. The vapor accelerates until it reaches the speed of sound. This causes the flowing vapor to compress itself by forming a high pressure

wave front as the vapor tries to flow faster than the speed of sound. This is the first stage of compression of the off-gas.

- The Diverging Section – The vapor slows down in a piece of pipe which gradually gets wider. The velocity of the vapor, as it slows, is converted to pressure. This is the second stage of compression of the off-gas.

The converging section and the diverging section are a team, working together, to boost the pressure of the off-gas. This two-man team is driven by the steam nozzle. The greater the energy of the steam delivered to the steam nozzle, the more energetically this two-man team will work.

Don't worry if you don't quite understand what I've just written. I'm not so sure I understand it myself. The entire subject of troubleshooting steam jets, surface condensers, seal drums, and barometric drain legs and their various malfunctions is somewhat of a mystery that I've been trying to solve for 47 years. So I'll try to illustrate this explanation again, several times, in growing detail, in the subsequent chapters of this book. But first, let me tell you a true story as an example of the complexity of this problem.

3.1 Tidal Flop in Delaware

Place: Delaware City Refinery (formerly Texaco)
Time: August, 2011
Subject: Tidal Flop
Client: Joy Hanson

On May 27th, 2011, I had brain surgery. I had taken a nasty fall a few months before, while running, and had hit my head. A hematoma (accumulation of fluid between the brain and skull) had gradually expanded and severely compressed the left hemisphere of my brain. Dr. Corales, my brain surgeon, explained:

"Mr. Lieberman, immediate surgery is required. Your left cortex has been compressed in excess of 30 percent, and blood flow is severely restricted."

"I kind of feel okay, doctor, except for a headache," I said. "I'm going to teach a refinery vacuum tower troubleshooting seminar next week. How about next month? I'll have more time then."

"Mrs. Lieberman," Dr. Corales said, turning to my wife, Liz, "Norman is close to death. He won't survive another month."

So, the surgeon drilled two, ¾" holes in my skull and drained out half a pint of fluid. This job in Delaware was my first refinery troubleshooting project since my operation. As I'm also an old guy, I wondered if my brain would still function. The problem at Delaware City, as described by Joy, the tech service manager, was the mysterious loss of vacuum occurring twice a day in their vacuum tower. Joy Hanson is exactly the type of gal I like to work with: a hands-on practical chemical engineer, interested in training her young contact engineers, able to delegate, but still insisting on involvement in solving malfunctions.

"Look Norm," Joy explained, "Here's the problem. Historically, our vacuum at the top of our vacuum column was four to six mm Hg. Now it runs ten to twelve mm Hg."

"That's really a tremendous loss in vacuum. I bet you're dropping 1,000 BSD of cat feed (Heavy Vacuum Gas Oil - HVGO) into your fluid coker feed (vacuum residue)," I observed. "That's like \$15,000 – \$20,000 per day you're losing."

"Yeah, Norm, but that's not our real issue," said Joy. "Twice a day, shortly after low tide on the Delaware River, the vacuum gets much worse. At low tide, our cooling water from the river gets 3°F to 5°F warmer. Then the vacuum creeps up to 15 or 16 mm Hg."

"That's real bad," I observed.

"Oh that's nothing," Joy Hanson exclaimed. "Sometimes the vacuum goes to 50 mm Hg or even 4 psia (i.e., 207 mm Hg). Then we lose the level in the LVGO draw-off (i.e., light vacuum gas oil) chimney tray (see Figure 3.1). The LVGO pump-around pump cavitates and then all hell breaks loose. We call this the tidal flop. This happens twice a day with low tide on the Delaware River. Mr. Collins, the new plant manager, is furious. It's upsetting the entire plant. Mostly, vacuum jumps up to 16 mm Hg or so, but sometimes, it is much worse."

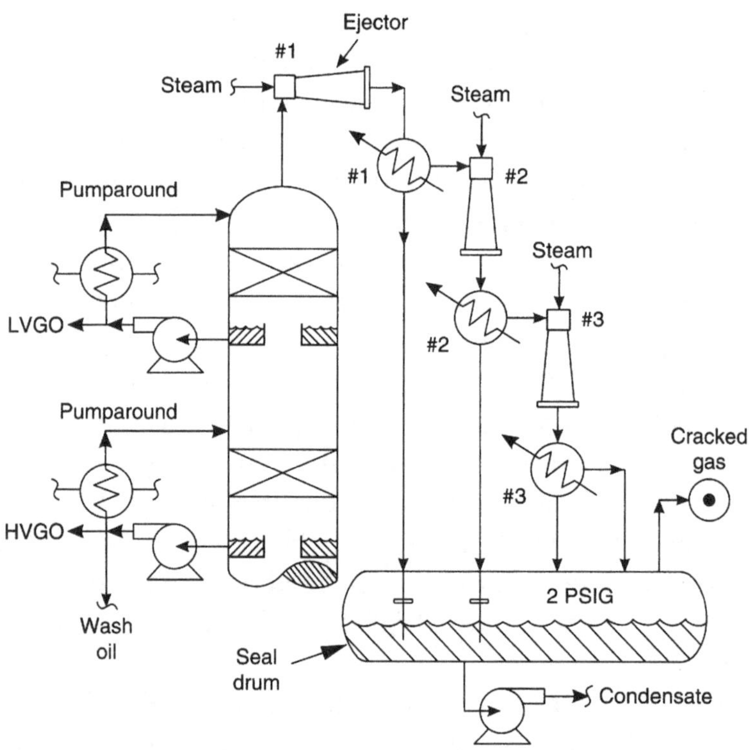

Figure 3.1 A three-stage jet system for a typical refinery vacuum tower. Most vacuum towers use this arrangement.

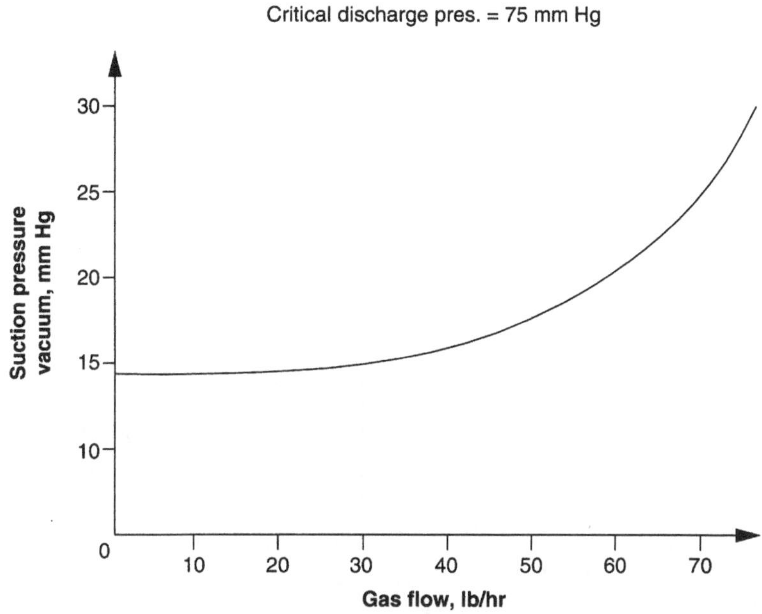

Figure 3.2 A typical vacuum steam jet performance curve.

3.2 Critical Discharge Pressure

I've shown a typical jet performance curve in Figure 3.2. The jet performance curve will indicate its design or critical discharge pressure. Once a jet's discharge pressure exceeds its design discharge pressure (also called its critical discharge pressure), then the jet is forced out of its critical mode of operation. The critical discharge pressure shown on this curve is 75 mm Hg.

That means the ejector loses its sonic boost or about half of its compression ratio. In this refinery, the critical discharge pressure was being exceeded because of an excessive pressure drop in the final condenser.

3.3 Fouling in Final Condenser

The discharge from the 3rd stage ejector, shown in Figure 3.1, flows into the top of the final condenser. This condenser is often

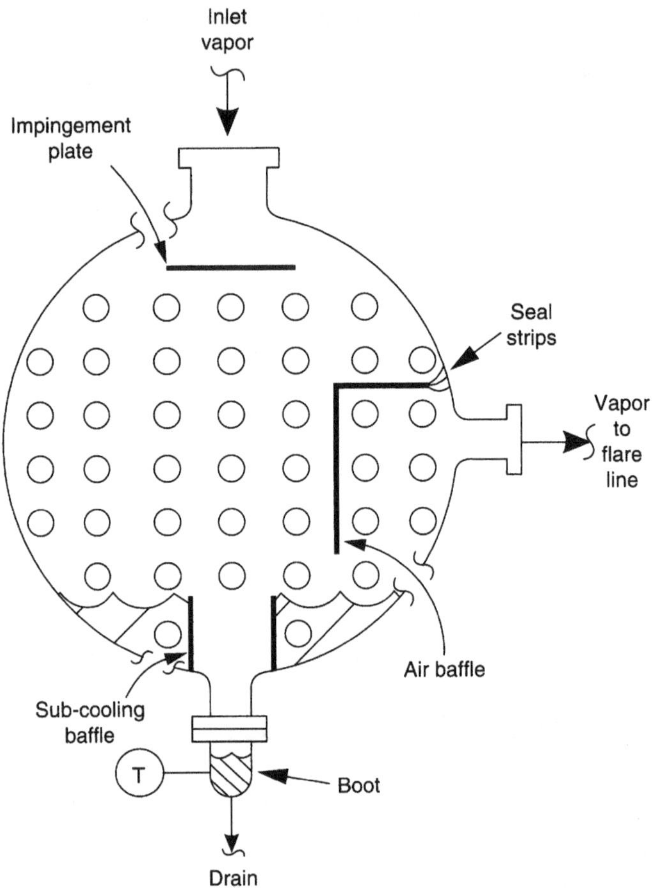

Figure 3.3 Internal components of a vacuum condenser. Seal strips retard leakage around air baffle.

subject to severe shell-side fouling, high pressure drop and low heat transfer coefficient. In the Delaware City Refinery, the final (#3) condenser delta P was not the typical 5 mm Hg shell-side, but 50 mm Hg. Of even greater significance, was that the boot on the condenser (see Figure 3.3) was partially plugged. Hence, liquid overflowed through the vapor outlet nozzle. The liquid (mainly steam condensate) accumulated in the large diameter vapor outlet line to the flare.

The problem was a four foot vertical section of line (i.e., a riser) between the 3rd stage condenser and a flare knock-out drum. It periodically filled with water. Four feet of water equals about 90 mm Hg. This hydrostatic head, plus the excessive 3rd stage condenser delta P of 50 mm Hg, was sufficient to cause the first stage ejector's critical discharge pressure to be exceeded. Thus, the #1 ejector lost its sonic boost and half its compression ratio.

3.4 Reduction in Back Pressure

I did two things to reduce this back pressure malfunction:

- First, I had the operators throttle the steam pressure to the 3rd stage ejector (see Figure 3.1). While the motive steam pressure to this jet was the design 160 psig, its downstream condenser limitations suggested that a lower than design motive steam pressure could be optimum. By trial and error, I adjusted this steam pressure to minimize the upstream pressure so as to optimize the upstream vacuum.
- Second, I had a hose connected from the bottom of the four foot riser pipe to the flare drum, to drain the steam condensate directly to the bottom of the flare knock-out drum. This eliminated the troublesome 90 mm Hg of excessive backpressure due to the four foot riser pipe, as I calculated just above.

3.5 Loss of LVGO Pan Level

The reader will now recall that the vacuum tower top pressure would exponentially increase to over 200 mm Hg during some of the "tidal flops", which I have yet to explain.

Let's assume that the increase in cooling water temperature associated with the tidal flop happened to coincide with the build-up of steam condensate in the four foot riser from the final condenser vapor outlet line. Then there would be a rather rapid increase in the vacuum tower top pressure. But how, I thought, could this lead to a loss of the liquid level in the LVGO pan level shown in Figure 3.1?

What a great moment this was to see whether Dr. Corales, my brain surgeon, had done a good job during my operation. Would my brain still function in its old way? Could I still unravel complex process engineering problems as I once did? It is the left hemisphere of the cortex, the part that engages in the analytical analysis, which is required for complex problem-solving that had been squeezed and compressed by my hematoma.

I already knew that the LVGO trap-out chimney tray was leaking at reduced vapor rates. A small increase in the HVGO pumparound heat extraction (Figure 3.1), which reduced the vapor flow to the LVGO pan, would cause a disproportionate drop in the LVGO product rate. That is, a substantial vapor flow through the LVGO chimney tray was needed to minimize tray deck leakage.

If the vacuum tower top pressure rapidly increased due to an increased discharge pressure from the #1 ejector, then the vapor flow to the LVGO chimney tray might easily be reduced by a factor of two or three, and then:

- The level in the LVGO chimney tray would disappear as the LVGO product leaked into the HVGO pumparound section.
- The LVGO pumparound pump would lose suction pressure and cavitate.
- The LVGO pumparound rate would become erratically low and heat extraction in the vacuum tower top pumparound would diminish.

- The uncondensed LVGO would flash up the tower and further bog down the #1 ejector, thus further increasing the vacuum tower top pressure.

To avoid the loss in LVGO pan level, I had to mitigate the effect of the increased vacuum tower top pressure. This, I did by making sure that as the cooling water temperature increased as a result of the changing tide in the Delaware River, that the four foot riser from the final condenser, was kept well drained. Also, I noted that the steam pressure to the primary jet was only 122 psig, as compared to the design minimum of 150 psig. By increasing this motive steam pressure, I enhanced the capacity and compression ratio of the #1 ejector and reduced its sensitivity to the increased back-up pressure.

The happy result of these changes was the complete avoidance of the loss of LVGO pan level due to tidal changes. Finally, this eliminated the large losses in vacuum that was upsetting the Delaware City Refinery vacuum tower.

But one problem still remained. What was the correlation between the increased cooling water temperature during the tidal changes and the persistent losses of five mm Hg experienced by the vacuum tower? The answer, as I'll explain, was due to the change to the vacuum tower feed composition.

3.6 Variations in Cooling Water Temperature

Let's refer again to Figure 3.1. In this refinery, the cooling water was drawn from the Delaware River tidal estuary. At low tide, the cooling water temperature in the summer would rise from 80°F to 85°F in about 30 minutes. Vacuum tower top temperature would increase from 100°F to 160°F, and then slip back down after the water temperature dropped to 80°F.

My first thought was that the first stage condenser, downstream of the primary ejector, would operate at a higher pressure, due to the warming cooling water. This would put a greater back pressure on this discharge of the primary jet, and raise its suction pressure. Indeed, I did measure that this condenser inlet pressure increased from 51 mm Hg to 54 mm Hg. However, 3 mm Hg primary ejector increased discharge pressure should only have increased the jet's inlet pressure by, at most, one mm Hg, not by six mm Hg. Also, the primary jet's critical discharge pressure was 75 mm Hg. As long as an ejector's discharge pressure is below its critical discharge pressure, small increases in its discharge pressure should have little effect on its inlet pressure.

Another confusing factor was that there was a substantial time lag of about twenty minutes between the time of the low tide effect on the cooling water temperature and the loss in vacuum. After studying the problem for a few hours, I made the following observations:

- As the cooling water temperature increased, the crude tower reflux drum temperature (see Figure 3.4) in the upstream crude distillation tower also became warmer.
- The increased reflux drum temperature increased the reflux drum pressure by about two psi. This gradually increased the pressure in the bottom of the crude distillation tower.
- I calculated that the amount of lighter boiling range hydrocarbons in the vacuum tower feed, flowing from the bottom of the crude tower, would increase by about 400 lbs/hr due to this extra two psi in the flash zone. In support of this calculation, a sample I drew from the bottom boot of the first stage condenser indicated an increase in its hydrocarbon content from 5% up to 8%.

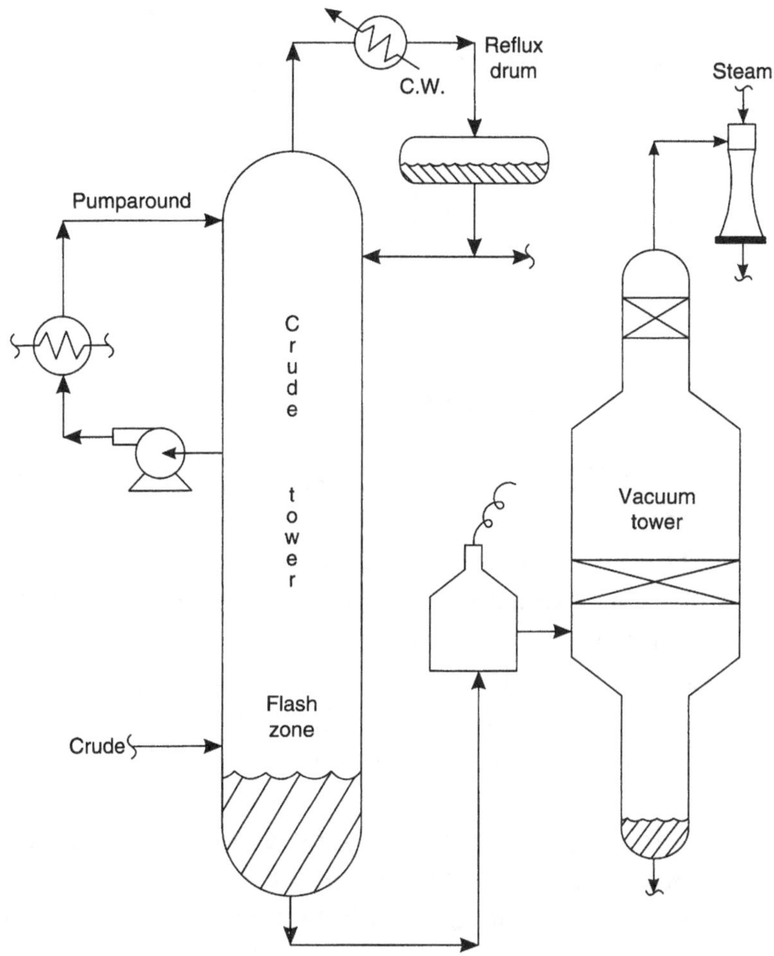

Figure 3.4 Higher reflux drum temperature causes a loss in vacuum.

> This corresponded to about 300 lbs/hr more hydrocarbon flow through the primary jet.

Looking at the ejector performance curve (Figure 3.2) supplied by the vendor, an increase of 400 lbs/hr of load to the primary condenser at a constant discharge pressure would result in an increase of 5 mm Hg to the first ejector. To prove my point, as well as to eliminate the loss in vacuum, during the next low tide, I raised the crude distillation tower top pumparound rate (Figure 3.4). This extracted more heat from the crude tower and prevented the increase in the reflux drum temperature, and hence the crude tower's bottom pressure also remained stable.

The result was that no vacuum was lost at low tide, even though the cooling water temperature did increase by its normal 5°F. And best of all, my now well-drained brain still apparently functioned.

3.7 Multi-Component Malfunctions

One of the difficulties in troubleshooting vacuum system malfunctions, as illustrated by this story, is that very often there are two or more problems plaguing the system simultaneously. For example, a fouled condenser downstream of an ejector is forcing the jet to surge because the jet's discharge pressure is excessive. However, if the motive steam to the jet was not wet, the jet could tolerate the high discharge pressure without surging (i.e., losing its sonic boost).

An ejector steam nozzle may be eroded. However, the ejector could still develop an excellent eight to one compression ratio, were it not for an air leak in the diffuser body. Excessive cracked gas could be tolerated if only the waste gas burner was not plugging and developing a high back pressure.

As I sit on a plane to Houston to visit the BP Refinery in Texas City, I know their main problem with their crude tower vacuum system is a fouled first stage condenser which cannot be cleaned due to its fixed tubesheet configuration (see Chapter 9). However, my client has also told me that when the system was new in 2004, it did not work properly even then. So my main challenge will be to define the original malfunctions which still exist, and which will be largely overshadowed by the fouled condenser.

Most of the vacuum systems I have worked on – both for condensing turbines and process vacuum towers – have suffered from several malfunctions simultaneously. It's this feature of vacuum systems that make troubleshooting these vacuum systems such a challenge.

3.8 Partial Tabulation of Vacuum System Malfunctions

In various sections of this book, I've explained some, but not all, of the vacuum system malfunctions I have dealt with. This list is far from complete. Eventually, I am sure, you will be able to add to it from your own experiences:

- Erosion to the steam nozzles.
- Leaking steam nozzle backing plate.
- Holes in diffuser body.
- Wet motive steam.
- Low motive steam pressure.
- High motive steam pressure.
- Excessive superheat in motive steam.
- Excess cracked gas and H_2S.
- Air leaks in overhead system.
- Air leaks in heater transfer line.
- Mechanical shaft seal worn on steam turbine.
- Nitrogen contamination of motive steam to steam turbine.
- Bad seal strips in air baffles.
- Condenser shell-side fouling.
- Condenser tube (water side fouling).
- Diffuser installed backwards.
- Seal leg plugged with wax.
- Steam nozzle partially plugged with hardness deposits.
- Reverse steam flow from two jets working in parallel.

- Excessive shell-side delta P.
- Loss of sonic boost in ejector.
- Ejector surging.
- Excess pressure drop in waste gas burner.
- Entrainment of gas oil from top of the vacuum tower.
- Air leak on seal leg outside of the seal drum.
- Internal leak inside seal drum on seal leg flange.
- Biological corrosion of the internal seal legs inside seal drum.
- Excessive seal drum pressure.
- Excessive seal drum level.
- Sludge accumulation in seal drum.
- Inadequate clearance between bottom of seal leg and seal drum.
- Condensate pumps undersized.
- Level in surface condenser covering the bottom edge of the air baffle.
- Seal strips in condenser leaking.
- Seal strips in condenser corroded due to use of copper alloys.
- Steam nozzle freezing-up.
- Use of fixed tubesheets for surface condensers.
- Low cooling water supply pressure.
- High cooling water return pressure.
- High cooling water supply temperature.
- Condenser elevation too high for cooling water pressure.
- Barometric drain lines too short for fluid density and seal drum pressure.
- Lack of velocity steam in vacuum tower heater passes.
- Excessive heater outlet temperature.
- Vacuum tower boot level too high.
- Vacuum tower boot too hot.

- Lack of adequate steam stripping of vacuum tower feed.
- Level tap purges using nitrogen or natural gas.
- Vacuum tower top temperature too hot.
- High liquid level in condenser boot due to level control error.
- Air leak in mechanical seal of condensate pump.
- Air leak in condenser boot level control stilling well or gauge glass.
- Air leak on packing glands of gate valves.
- Vacuum tower feed contaminated with crude oil.
- Excessive vacuum tower bottoms stripping steam rates.
- Naphthenic Acid degradation to CO.

I have only stopped adding items to my list in the fear that I will frighten potential readers. And the worst part of this story is that I've been adding to my list for 47 years, and I am still adding to the list at an undiminished rate to this day.

What can I say? My best advice is not to get involved in troubleshooting vacuum system malfunctions in the first place. Stockbrokers and lawyers make the big bucks...not us.

However, if you have the misfortune to be involved in troubleshooting a vacuum system problem, my book is probably the best help you can get.

4

Effect of Water Partial Pressure on Jet Efficiency

"Mr. Lieberman, I've been informed that you opened a sample bleeder on our Benzene Stripper labeled *Toxic Vapors. Do Not Open,*" said the Exxon safety director at the Beaumont Refinery

"Yes, that's right," I answered.

"Mr. Lieberman, did you not see the sign prohibiting such an action?"

"Sure I saw it. It was right next to the valve I opened."

"But why then did you open it? Surely you don't think you know more about these matters than the Exxon Engineering Division?"

"Look," I explained, "I don't think I know more than Exxon Engineering. I'm sure that I know more than the entire Exxon Corporation."

4.1 Vapor Pressure of Water Limits Vacuum

Feeling I had better back up my boast, I drew Figure 4.1. That's the old ESSO LAGO #4 V.P.S. in Aruba. This giant vacuum tower had eight 60" I.D. pre-condensers, upstream of the first stage jets. The condensers were cooled by sea water, which is pretty constant in Aruba at about 80°F. The cooling water outlet temperature from all eight exchangers was about 90°F.

The performance of the vacuum system was limited by the vapor pressure of water at the pre-condenser vapor outlet temperature. Let me explain.

The vapor leaving the vacuum tower was mainly vacuum tower stripping steam. The main purpose of the eight large pre-condensers was to condense most of the steam. The uncondensed steam, along with the cracked off-gas, was drawn into the first stage jets. Most of the vapor load to these jets were moles of steam.

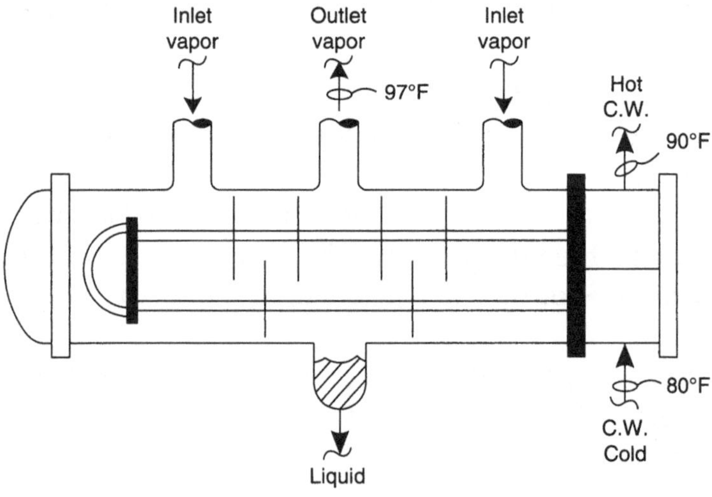

Figure 4.1 Cooling water flowing in the wrong direction increases precondenser outlet temperature.

Table 4.1 Vapor Pressure of Water in mm of Hg.

Temperature (°F)	Vapor Pressure Water mm of Hg
80°F	25
90°F	37
100°F	51
110°F	66
120°F	87

Here's the key point. At the 97°F (see Table 4.1) vapor outlet temperature, water has a vapor pressure of 48 mm Hg. I had measured a pressure at the condenser vapor outlet essentially at the same 48 mm Hg. Thus, the vacuum system performance was limited by the vapor pressure of water at the pre-condenser outlet (see Chapter 2). Improving the capacity of the downstream jets would only suck more moles of water out of the pre-condensers without improving vacuum.

Referring back to Figure 4.1. Note that Exxon had followed the normal convention of placing the cooling water inlet on the bottom of the channel head and the water outlet on the top of the channel head. Almost every water cooler I have ever seen is like this. However, note that both the vapor inlet and the dual vapor outlet nozzles are all on the top side of the shell. This is a characteristic of the split-flow condenser shown.

Finally, note that cooling water outlet of 90°F is only 7°F below the 97°F vapor outlet. This is called the hot end approach temperature. A 7°F value for this parameter is extremely small (that is, good).

This means that the best vacuum that can be produced is limited by the 90°F cooling water outlet temperature. That is, 37 mm of Hg as indicated by Table 4.1. And this gave me an idea. Suppose I was to reverse the cooling water flow. Then, the 80°F sea water could reduce the 97°F vapor outlet by another few degrees. This would suppress the vapor pressure of water to the first stage jets.

So, I had a meeting with Mr. English, Coastal's plant manager of the former Esso (Exxon) Lago Refinery in Aruba.

Mr. English looked at my calculations and my proposed revised cooling water piping sketches.

"Norman," he said, "Do you think you know more than Exxon Engineering?"

"No, sir," I answered. "I don't think I know more than the Exxon Engineering Division. I'm sure I know more than the entire Exxon Corporation."

"Okay. Norm, let's find out who is smarter – you or Exxon."

So the direction of the cooling water flow was reversed. The fresh cool sea water inlet was placed on the top of the channel head. The pre-condenser's outlet temperature dropped from an average of 97°F to 94°F, which reduced the vapor pressure of water by 3 mm Hg, and reduced the vacuum tower pressure by the same amount—3 mm of Hg. Gas oil recovery from the vacuum tower bottom's tar improved by 1,400 BSD.

I explained all this to the safety director of the Exxon refinery in Beaumont. His final words, however, were not encouraging.

"Mr. Lieberman, I am permanently barring you from this facility. May I now have your visitor I.D. pass?"

Incidentally, I have published in *The Oil & Gas Journal* the whole story about the benzene stripper fiasco. I didn't mention Exxon by name, but the management of the Exxon Refinery at Beaumont, Texas was still furious. The reference, if you're interested, is: O&GJ, July 18th, 1994, page 50, "Basic Technology Can Improve Refinery Profits," by Norman P. Lieberman.

As I write these very words, I'm on a plane from New Orleans to England, to teach a troubleshooting seminar at the Esso (Exxon Mobil) refinery in Fawley. I'll tell the young engineers my stories and they'll laugh. But in a way, it's all rather sad. Read my short article in *The Oil & Gas Journal* and you'll see my point.

4.2 Reminder about Water Partial Pressure

After you have identified all your various sources contributing to your off-gas flow, don't make the same mistake that I sometimes do. That is, I forget to back-calculate the moles of H_2O. Under vacuum conditions, the vapor pressure of water at the condenser vapor outlet temperature can easily be a very large part of the total pressure. It is not at all uncommon for the partial pressure of H_2O to be 50% to 80% of the total pressure. Then, convert the moles of H_2O to its air equivalent. An accurate field measurement of the condenser vapor outlet temperature is quite critical in this service to avoid misstating the moisture partial pressure.

4.3 Air Leaks in Steam Turbine Surface Condensers

Leakage of air into the surface condenser is a serious matter. A small leak could easily load-up the ejectors. To see if jet capacity and/or air leaks are limiting the surface condenser vacuum, you should proceed as follows:

- Step One – Check the temperature at the vapor outlet of the condenser (not the liquid outlet or boot temperature).
- Step Two – Compare the measured pressure at the surface condenser vapor outlet to the vapor pressure of water (from your steam tables or from Table 4.1), at the temperature determined in Step One.

- Step Three – If the measured pressure is more than five mm of Hg greater than the calculated vapor pressure of water in Step Two, the limiting factor for the surface condenser performance, is a combination of an air leak and a lack of required ejector capacity.

For vacuum tower surface condensers, the above limitation also applies, but then the problem is not only air leaks, but the flow of cracked gas as well.

4.4 Variable Cooling Water Temperature

Big Springs is located in the scrub desert of western Texas. Cold autumn mornings are followed by warm afternoons. The local refinery's cooling-water temperature follows this ambient temperature trend. The vacuum tower in this refinery also seems to keep track of the time of the day.

At 6:00 a.m., the primary steam jet is running quietly and pulling a vacuum of 12 mm Hg. At 7:30 a.m., the jet begins to make infrequent surging sounds. It rather sounds as though the jet is slipping every 15 or 20 seconds. As the morning coolness fades, the surging becomes more frequent. The vacuum also begins to slip from 12 to 14 mm Hg. Then, about 9:00 a.m., as the surges have become so frequent as to be almost continuous, the vacuum plunges to 23 mm Hg. By 10:00 a.m., surging has stopped and the vacuum at the jet suction has stabilized at a poor 25 mm Hg. Many, if not most, operators of large vacuum jet systems have observed this problem – but what causes it?

4.5 Loss of Sonic Boost

Let's refer to Figure 4.2. As the cooling water warms, the vapor outlet temperature of condenser No. 1 increases. This

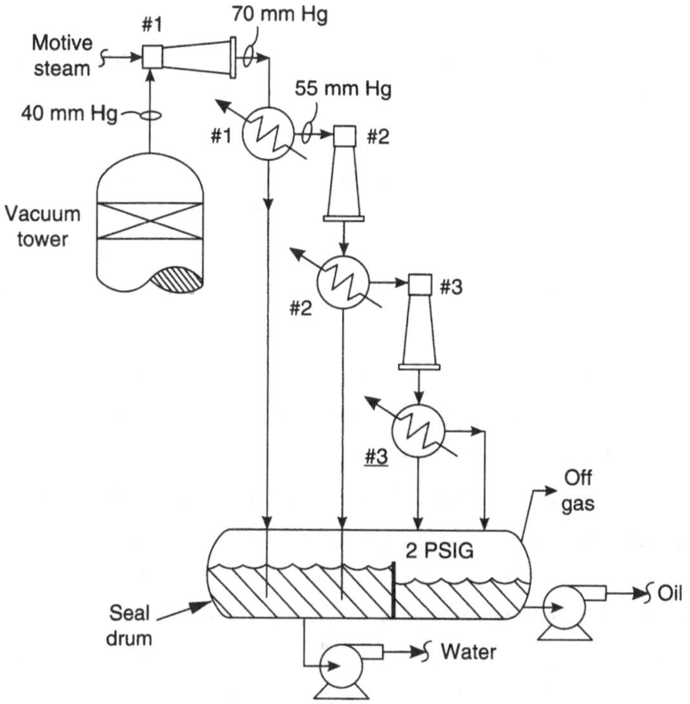

Figure 4.2 Increased cooling water temperature caused vaccum to break by exceeding the critical discharge pressure of the stage ejector.

also increases the condensing pressure due to the higher vapor pressure of water, as shown in Table 4.1. This raises the discharge pressure of the primary jet (jet 1), as well as the pressure in the diffuser throat (see Figure 1.1). Higher pressure will result in a smaller vapor volume. And a smaller vapor volume will cause a reduction of the velocity in the diffuser throat.

The lower velocity in the diffuser throat does not affect the jet's performance, as long as the velocity remains above the speed of sound. If the velocity in the throat falls below the speed of sound, we say that the jet has been forced out of critical flow. The sonic pressure boost is lost. As soon as the sonic boost is lost, the pressure in the vacuum tower suddenly increases. This partly suppresses vapor flow from the vacuum tower. The reduced vapor flow slightly unloads condenser 1 and jet 2, shown in Figure 4.2. This briefly draws down the discharge pressure from jet 1. The pressure in the diffuser throat declines. The diffuser throat velocity increases back to, or above, sonic velocity. Critical flow is restored, and so is the sonic boost. The compression ratio of the jet is restored, and the vacuum tower pressure is pulled down. This sucks more vapor out of the vacuum tower, and increases the loads on condenser 1 and jet 2 (the secondary jet). The cycle is then repeated. Each of these cycles corresponds to the surging sound of the jet and the loss of its sonic boost.

As the cooling-water temperature rises, the vapor pressure of water also rises. The sonic boost is lost more easily and more rapidly. The surging cycles increase in frequency to 30 or 40 per minute. The vacuum tower pressure becomes higher and higher. Finally, the surges become so frequent that they blend together and disappear. The primary jet has now been totally forced out of critical flow. The sonic boost has been lost until the sun in Big Springs, Texas, sets and the desert cools. Surging then returns until the critical flow in the jet is restored as the vapor pressure of water declines in the condenser, and the sonic boost is regained at about 9:00 p.m. in the evening.

4.6 Relative Jet Efficiency

Recently, I was working at a refinery in Coffeyville, Kansas. The operators had been running two jets in parallel. They wished to take one of the two jets out of service. I checked the mixing section skin temperature on both jets. Basically, I was measuring the exit temperature of the steam nozzles (see Figure 2.1):

- Jet A = 90°F
- Jet B = 110°F

Which jet is the least efficient and thus should be removed from service? As you consider this question, please note:

- The heat content or enthalpy of the steam is converted to velocity in the steam nozzle (see Chapter 1). The more kinetic energy the motive steam has, the better the jet will work in compressing the off-gas from the vacuum tower.
- Both jets were supplied with the same quality (moisture content) motive steam.

The answer is Jet "B," because the steam is warmer and hence presumed to be moving slower. If the steam supply to both jets was different, I would most likely have thought Jet "A" suffered from poor quality (that is, high moisture content) steam.

4.7 Definition of "Vacuum Breaking"

When an operator says the vacuum has broken, 90% of the time he's referring to the first stage jet having lost its sonic boost. Most recently, I was working in a refinery where the tower top pressure would increase from 6 mm Hg to 15 mm Hg over perhaps a five to ten minute interval. Since the jet had a constant discharge pressure of about 40 mm Hg, the jet's compression ratio decreased from 6.6 (i.e., 40 ÷ 6) to 2.6 (i.e., 40 ÷ 15). That's typical. In this book, when I use the term, "loss of sonic boost," I could also say, "The jet has broken."

But sometimes, a vacuum tower or surface condenser will have a sudden increase in pressure unrelated to jet performance. A few examples are:

- The liquid level in the pre-condenser rises above the bottom lip of the internal air baffle.

- There is a sudden increase in the gas flow to the first stage jet.
- The seal drum liquid level is lost.
- The seal drum floods.
- The LVGO pumparound floods.

Why then, do I specify that the vacuum breaking is due to loss of the sonic boost in the first jet? Because if the sonic boost is lost in a downstream ejector, it will result in the first stage jet discharge pressure exceeding its critical discharge pressure, thus causing it also to lose its sonic boost compression.

4.8 Critical Discharge Pressure Exceeded

I just got off the phone with my pal, Jerry. We've been discussing the vacuum system at the Citgo Refinery in Corpus Christi, Texas (see Figure 4.2).

Jerry noted that the vacuum tower top pressure was about 40 mm Hg. The discharge pressure from the jet was 70 mm Hg and the pressure at the outlet of the No. 1 condenser was 55 mm Hg, for a No. 1 condenser delta P of 15 mm Hg. The design condenser delta P was only 5 mm Hg. The design discharge pressure from the primary jet was 60 mm Hg.

Note that the actual compression ratio of the primary jet was less than two to one (70 ÷ 40 mm Hg). As per the jet performance curves, the design compression ratio at the current measured flows was four to one (60 ÷ 15 mm Hg).

The 70 mm Hg actual jet discharge pressure exceeded the design or critical discharge pressure of the first stage jet, which was 60 mm Hg. When this happens, a converging-diverging jet will be unable to develop its design compression ratio of four to one, and can only develop a compression ratio of two to one.

The vendors say the jet has been forced out of its critical mode of operation. I say, the jet has lost its "Sonic Boost." (See

Chapter 2 of this text.) The vendors and I both mean the same thing. The jet discharge pressure is above its critical or design pressure.

Thus, I had told Jerry to wait for the next cold front that came through in the evening. This would, at least for a few hours, suppress the discharge pressure from the first stage ejectors by lowering the condensing temperature and pressure in the first stage or primary condenser.

The next morning, Jerry phoned me to transmit the data observed the previous night at Corpus Christi:

- Inlet pressure to the first stage condenser dropped 10 mm Hg from 70 to 60 mm Hg (see Figure 4.2).
- Vacuum tower top pressure dropped from 40 mm Hg, down to 25 mm Hg.

Something was causing an excessive delta P in the No. 1 condenser, as the vapor loads were all below design. Also, the heat transfer coefficient in this condenser was also less than half of design. Typically, since this exchanger was just cleaned, I would guess an oversized impingement plate and bad seal strips along the top edge of the vapor or air baffle. I've discussed these types of problems in subsequent chapters in this text.

To summarize, any time a jet's design or critical discharge pressure is exceeded, the jet will lose its sonic boost, and thus only develop half of its design compression ratio. Don't take this statement as a hard and fast rule. To an extent, it depends on the motive steam moisture content, the physical condition of the motive steam inlet nozzle, and the vapor inlet flow rate.

I might also note that because my clients often do not have a fundamental grasp of their real problem, they will spend a small fortune in changing out the motive steam nozzles on all ejectors of their vacuum system.

5

Air Leaks

One of the questions I am sometimes asked is, can an air leak in a vacuum system cause an explosion or an internal fire?

I do have some indirect evidence that air will oxidize iron sulfide deposits in the downstream condenser. I've allowed air to enter upstream of a first stage condenser for five minutes thru a ½″ bleeder valve as an experiment and altered the operating characteristics of the condenser for several months thereafter. This was altered in an unfavorable sense, since the vapor distribution appeared to worsen and the condenser vapor outlet temperature increased by 5°F. However, no external damage to the condenser occurred.

5.1 Upper Explosive Limits

For the gas inside a vacuum condenser to explode, the concentration of the hydrocarbons would have to be less than about 15%. For example, the upper explosive limit for methane is around 15%. For propane, it's about 12%. Even a large air leak is unlikely to dilute the hydrocarbon content of vacuum tower off-gas to such a low level.

Certainly, in my 46 plus years of experience, I am unaware of any explosions or temperature excursions in the vacuum systems serving refinery vacuum towers. I have often opened bleeders to draw air into vacuum systems to gauge the effect of an increased non-condensible vapor load on the ejector capacity. I have observed nitrogen contents of up to 50% in the seal drum off-gas. But, I would suggest above this air leakage rate (50% N_2), I might begin to worry about the formation of a combustible mixture and the remote possibility of an explosion.

5.2 How to Find Air Leaks

When air expands, it cools. All gases (other than H_2 and CO_2) cool upon expansion. Atmospheric air, when drawn through a small hole into a vacuum system, will cool. Not only because the air has a positive Joule-Thompson Expansion Coefficient, but also because a lot of its enthalpy (heat) is being converted into kinetic energy (speed). This will produce a cold spot on a leaking pipe or fitting. In New Orleans, the land of humidity, this is going to cause a localized condensation to form on the pipe. If you touch such a spot and it's cold, there is an air leak.

To confirm your air leak, I would wrap it with tape. If it's the sort of leak that can't be wrapped (like a screwed fitting), or the packing gland of a valve, pour water over the suspect leak. For steam turbine surface condensers, you will instantly see if the air flow from the atmospheric vent diminishes. Within a few minutes, vacuum would improve if your ejectors were limiting the performance of the surface condenser.

Not uncommonly, the air leak could be in a flange. Wrap the suspect flange with tape. Punch a hole in the tape and spray aerosol shaving cream over the hole in the tape. If the cream is sucked in, there is an air leak.

A more expensive method (which I have never used, but heard about from my wealthy clients) is:

- Spray Helium gas over a suspected air leak area.
- Install a portable He monitor on the non-condensable vent.

You can do the same thing yourself without hiring an expensive contractor. Use air conditioner Freon and a portable Freon leak detector available at your local air conditioner supply store. This is a good general method to detect leaks in many heat exchangers as well.

5.3 Diffuser Air Leaks

The diffuser (see Figure 2.1) is by far the largest mechanical portion of a steam ejector. While its interior is subject to erosion, I can say from my experience, that this does not matter very much. A little bit of erosion to the steam nozzle can degrade an ejector's performance, but moderate erosion does not affect the diffuser.

However, in refinery vacuum towers, the diffusers are subject to HCl erosion-corrosion, which if left unchecked will eat a hole right through the body of the diffuser. Then, air will be sucked into the jet and ruin its performance. I've never experienced this problem myself, as most of my clients inject NH_3 or neutralizing amine into the steam supply to their first stage jet to control HCl corrosion. But, I have seen how my clients repair such damage. There are two alternatives, both of which seem to work equally well:

1. Shut-down and disassemble the ejector and send the giant diffuser body (sometimes thirty feet long) back to the manufacturer for repair or replacement.
2. Wrap aluminum duct tape around the leak – which appears to last for years.

I vote for option number two.

5.4 Air Leaks on Vacuum Towers

Finding air leaks in refinery vacuum systems is more difficult than in steam turbine exhaust surface condensers. Not only is the system far bigger, but the air leaks could be associated with the hotter parts of the system. So you will need to get a sample of off-gas in a non-metallic container. This is critical. You must prevent the reaction:

- $H_2S + O_2 \rightarrow H_2O$ + sulfur

Metal surfaces catalyze this reaction.

The O_2 content must be preserved. For example, let us assume that you obtain a sample of the vacuum tower tail gas. It's 40% N_2. You would suspect that you would then have 11% O_2:

- (40%) • (21% ÷ 78%) = 11% O_2

But you don't. You have 1% O_2 and 10% CO_2! What happened to the O_2? And where did the 10% CO_2 come from? Well, the air leak was in the vacuum heater transfer line, which ordinarily runs at a slight sub-atmospheric pressure. In the Coastal Refinery in Aruba, we once found a hole in the last vacuum heater tube, based on the CO_2 and N_2 content of the vacuum jet off-gas. That is, the O_2 was burning the hydrocarbon content of the feed, and thus producing CO_2. Of course, some of the O_2 is converted to

CO and H_2O. So my main point is, if the O_2 content is disproportionately low, compared to the N_2 content, then the air leak is in the hot part of the vacuum tower or heater, and not in the overhead system.

Incidentally, if you have the following sort of analysis of your non-condensables:

- $N_2 = 20\%$
- $O_2 = 5\%$
- $CO_2 = 0$
- $CO = 5\%$

The Carbon Monoxide content is not due to an air leak, but is caused by the thermal degradation of Naphthenic Acids. There is nothing you can do about this reaction, so don't worry about it.

5.5 Air Leaks in Heater Transfer Lines

This is a safety issue of some importance. I have been involved in the apparent sudden failure of vacuum heater transfer lines at the American Oil Refinery at Sugar Creek, Missouri and at the Good Hope Refinery in New Orleans. I say "apparent" sudden failure because the leaks in the heater transfer line had not developed suddenly at all. If we had been tracking the N_2 and CO_2 content of the off-gas, we could have seen the leak developing for months. It's just that this leak was under the insulation covering in an area of the heater transfer line operating under a slight sub-atmospheric pressure. An older student from Exxon in one of my troubleshooting seminars said he once removed the insulation from a vacuum heater transfer line and could see the feed rushing past him. We found the leaks at Sugar Creek and in Louisiana in a more dramatic fashion. A minor problem with the steam jet systems caused a sudden loss of the tower vacuum. The defective section of the heater transfer line went to a positive pressure. The hot oil blew out, auto-ignited, and large fires resulted.

Why are transfer line leaks so common (transfer line is the term used to describe the pipe connecting the heater to the tower) in vacuum services?

- Naphthenic Acids inherent in many crudes.
- High (750°F – 800°F) temperature.
- But mainly, high velocity. Especially if the velocity exceeds 50% of sonic velocity, and the transfer line is constructed (God forbid) out of 410 chrome steel. I believe 347 stainless is the preferred metallurgy for this service.

So, the gradual increase in CO_2 in off-gas, even if it's not greatly affecting vacuum, should still be considered as an indication of a malfunction that you should worry about.

5.6 Air Leaks – Turbine Mechanical Seal

For an overloaded vacuum system serving a steam turbine, there is only one possibility. That is an air leak. Most of the normal air in-leakage occurs along the turbine's shaft seal. When the turbine is not running, air leakage along the shaft, which is drawn into the surface condenser, is normal. But, when the turbine starts to spin, and the shaft begins to get hotter, most of this seal leak stops. My observation is based on the vacuum in the surface condenser improving, once the turbine is up and running.

To a large extent, the air leak in the turbine's mechanical seal should be extracted by the small hogging type steam jet on the side of the turbine, located near the coupling (see Chapter 18, "Other Types of Vacuum Systems"). Thus, the small steam jets located downstream of the surface condenser should ideally only be extracting the air leaks in the surface condenser itself.

The ejected air is exhausted out of a vent discharging from the final (second) stage condenser. If there are no air leaks, there will be no flow from this vent. It is not at all unusual for me to set a piece of cardboard on top of this vent, without changing the pressure in the surface condenser which is a certain indication of no air leaks. On the other hand, you can hold a strip of cloth over this vent and get a qualitative idea as to how much air is being exhausted. Then, as I reduce the air in-leakage, I can easily observe if the atmospheric vent is blowing with reduced velocity.

6

Sources and Disposal of Hydrocarbon Off-Gas

The rate of formation of cracked gas is a function of time and temperature. By time, I mean the residence time that liquid is held at a particular temperature. The proper way of stating this, is that thermal cracking is a Zero Order Reaction. In Figure 6.1, I have presented a chart correlating:

- Observed Gas Flow from Jets
- Barrels of Vacuum Tower Feed
- Vacuum Tower Flash Zone Temperature

6.1 Evolution of Cracked Gas

This chart is based on very old American Oil empirical observations, and thus is a useful design tool. But this chart and the theory of thermal cracking being a zero order reaction is suspect. The problem is that cracked gas can contain 40% H_2S

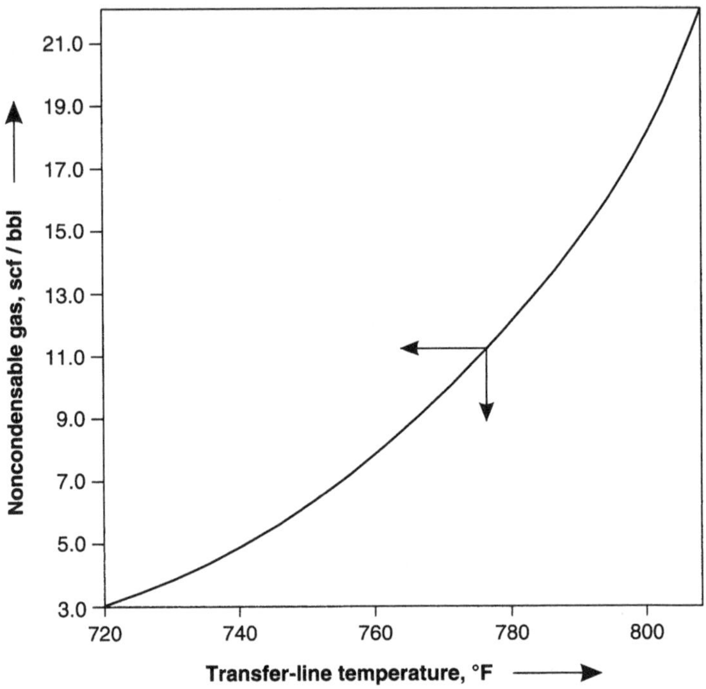

Figure 6.1 Heater outlet temerature vs cracked gas rate. Typical maximum is 770°F to 780°F. Excludes allowance for air leaks and cracked gas formation in boot.

on a dry basis. Hence, higher sulfur feeds will generate more dry-gas than low sulfur feeds.

The term Dry Basis means that I have condensed out the moisture in my sample of off-gas before checking the H_2S content with my Drager Tube or Gas-Tec Tube. So, before you kill yourself with H_2S (fatal concentration being 1,000 ppm or 0.1% in air), here's how to check for H_2S and safely obtain a sample of gas from a seal drum or hot well:

- The hot well will have a slight positive pressure of one to four psig.
- Put on your Scot Air Pak or B.A. equipment.
- Get a pint size can or bottle and two lengths of plastic tubing.

- Blow gas into your bottle from the bleeder on the off-gas line thru a plastic tube, from the hot well drum. Open the valve just enough so that you can feel a little gas blowing out of your bottle.
- Draw the sample up into your glass drager tube (this is a standard, portable, lab tool, used to measure, by chemical reaction, the concentration of various gases such as H_2S, SO_2, CO_2, NH_3, etc., on a dry basis).
- To retain a sample for the lab, I inflate a bicycle tire inner tube using a hand pump (which looks like a rubber ball).

In Aruba where the vacuum tower feed might be 4 wt.% sulfur, I measured 40% H_2S in such a sample of cracked gas. In other places, with the sulfur content of vacuum tower feed is about one percent, I've measured about 15% H_2S in the off-gas from the seal drum. So obviously, a large amount of the cracked gas rate is related to the sulfur composition of the feed. Hence, evolution of cracked gas is actually dependent on things other than time and temperature. Some engineering companies simply assume one pound of cracked gas per barrel of vacuum tower feed for a heater outlet of 780°F.

In order of concentration (on a dry basis), a typical off-gas from the vacuum tower seal drum would contain:

- Methane (up to 50%)
- H_2S (10% to 40%)
- Hydrogen (maybe 10%)
- C_2's (maybe 10%)
- C_3's and C_4's (maybe 10%)
- Nitrogen (about 5%)
- CO_2 and CO (about 2%)

This analysis excludes air leaks. The 5% N_2 is due to unavoidable air contamination on sampling. A typical molecular weight is 30

6.2 Sources of Cracked Gas

There are three areas where cracked gas is generated. All three are under our direct control:

- Vacuum Heater Passes or Tubes
- Transfer Line and Flash Zone
- Bottom Boot of the Vacuum Tower

The vacuum tower heater operates differently than other process heaters. That is, the maximum temperature that is experienced in the heater is not at the heater outlet, but back upstream in the heater passes, which is perhaps two, three, four or more tubes upstream of the outlet. That's because as the flow nears the outlet, it undergoes a big delta P. A lot of flashing occurs as the heater outlet pressure falls to the sub-atmospheric pressure near or at the outlet. All this evaporation or flashing requires lots of latent heat of evaporation. Some of this heat does come from the radiant heat transfer in the tube. But, most of it comes from the flowing oil itself. That is, the rapid reduction in pressure causes the sensible heat content of the flowing oil to be converted to latent heat of evaporation. Thus, the oil temperature drops, even though radiant heat transfer continues. Hence, the heater outlet temperature might be 40°F–50°F lower than the peak tube temperature. This value of 40°F or so, is not a calculated number. At the Mobil Refinery in Coryton, U.K., I observed just such an effect on their vacuum heater, which had many TI points through the heater passes.

Steam in the heater passes reduces the hydrocarbon partial pressure. This promotes the earlier vaporization of the gas oil in the heater passes and thus suppresses the peak temperature. The rate of cracked gas evolution is strongly related to this peak temperature, rather than to the heater outlet temperature, which is cooler.

Unfortunately, this bit of knowledge once cost me $65,000. You see, I had a contract from the Unocal Refinery in San Francisco, to revamp their vacuum tower for increased gas oil recovery. Rather than just carry out my assignment like a sane and sensible engineer, I decided to run a plant test.

- I increased the steam in the heater passes from zero to 1½ pounds of steam per barrel of vacuum heater feed. The heater inlet pressure increased from 60 psig to 100 psig. I would have used even more steam, but the heater inlet pass valves were almost wide open.
- The cracked gas flow went down a lot.
- The vacuum became better as I unloaded the jets.
- I gradually increased the heater outlet temperature until the vacuum started to slip back down.
- Next, I optimized the steam pressure to the jets by trial and error.
- Then, I cranked up the cooling water flow to the vacuum tower overhead pre-condenser, and added a little more steam to the vacuum tower bottoms stripping section.
- Next, I added a few more pounds of steam to the heater passes and a few more degrees to the heater outlet and back-flushed the vacuum tower pre-condenser.
- And then gas oil production increased to the required design value of 9,000 BSD. And then my former friend, Paul Schroder, canceled my $65,000 contract and issued a $2,000 P.O. number for my one day's field services. Finally, I remembered what my Dad used to tell me: "Son, you are selling knowledge, not time. Go dig ditches if you want to sell time."

6.3 Cracked Gas Evolution from Boot

Please refer to Figure 6.2. Large amounts of cracked gas can be evolved from the vacuum resid in the tower bottoms. On one unit at Murphy Oil in New Orleans, we found that 50% plus, of the evolved gas was due to cracked gas from the bottom's boot. You can see if this is a problem by lowering the boot quench temperature. Or, if you do not have such quench facilities, reduce the tower bottoms level.

If the cracked gas rate goes down, then your vacuum is going to get better. Typically, keeping the tower bottom's temperature below 680°F avoids the whole problem. But, if the boot's residence time is low, boot temperatures above 700°F are also okay. I've noticed lately that bitumen from Canadian tar sands can require even lower boot temperatures.

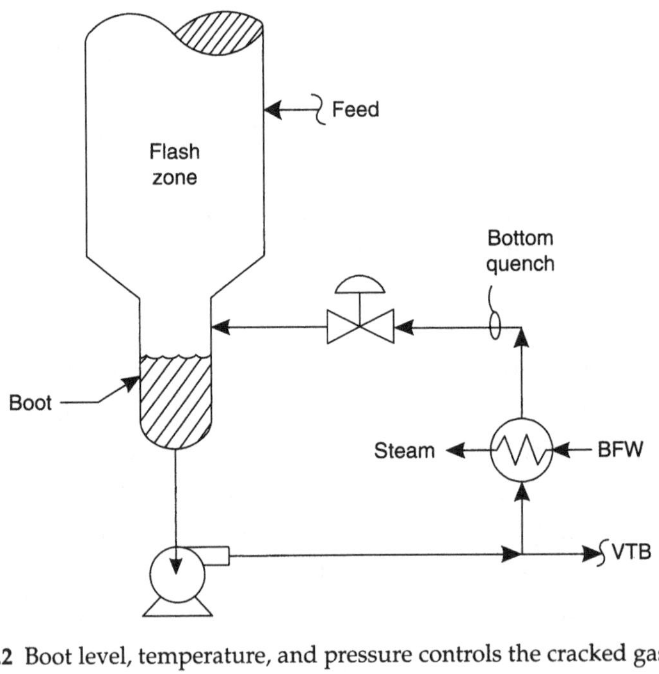

Figure 6.2 Boot level, temperature, and pressure controls the cracked gas evolution rate.

6.4 Air Equivalent

Some performance curves will be tabulated in "air equivalent" and some will be "steam equivalent". There is a rigorous method to convert different molecular weight hydrocarbons to either air or steam equivalent. Roughly speaking, you take the square root of the ratio of the molecular weight. For example, let's say I have 100 pounds per hour of butane vapors of molecular weight of 58. Air has a molecular weight of half of 58. That is 29. The square root of 58 ÷ 29 = 1.414. Therefore, I would divide my 100 pounds of butane by 1.414 to obtain an air equivalent flow of 100 ÷ 1.414 = 71 pounds of air.

For practice, see if you can calculate that ten pounds per hour of steam has an air equivalent of about 13 pounds per hour (square root of 29 ÷ 18 times ten pounds).

The composition of the gas is based on an off-gas sample obtained from the top of the seal drum, or the vapor outlet of the final condenser, or from the vapor flowing to the waste gas burner. It's best not to try to obtain the sample from a sub-atmospheric pressure point and risk air contamination. Take the sample carefully in a non-metallic container to avoid having tramp oxygen reacting with H_2S. Be careful! The H_2S concentration will be 100,000 to 300,000 ppm in crude unit vacuum tower seal drums.

To obtain the sample, use a quart bottle as described above.

But how about measuring the flow? Here's the method that I use to get the gas flow from the final condenser:

- Step One – Check the off-gas for nitrogen. Assume for this case, that the lab result is 5% nitrogen.
- Step Two – Connect a fitting, with a ¼″ to ½″ opening anywhere on the first stage condenser.
- Step Three – Open the valve to this fitting all the way. If the vacuum starts to break or the jets start to surge, use a fitting with a smaller opening.

- Step Four – Check the off-gas for nitrogen. Assume the lab result is 15%. You only need to wait two minutes between steps three and four.
- Step Five – Assume the velocity of air sucked through your fitting is 1,000 ft/second (i.e., about sonic velocity). Using the fitting open area, calculate the air flow, in ft^3 per second. For a ½ inch open fitting, multiply by 3,600 to get 5,000 SCF/HR.
- Step Six – Your sample has increased its nitrogen content from 5% to 15%. The nitrogen content of air is 79%. Therefore to calculate the flow of off-gas:
  - $(5{,}000\ ft^3/sec)\ (79\% \div (15\% - 5\%)) = 39{,}500$ SCF/HR
- Step Seven – To convert to moles per hour, divide by 360 to get 110 MPH.
- Step Eight – Now, you have to add in the moles of condensable hydrocarbon to the first stage jet. If there is a pre-condenser, then the moles of condensable hydrocarbons to this jet are small and can be neglected. If there is no pre-condenser, then continue on to Step Nine.
- Step Nine – Measure the volume of liquid hydrocarbons collected in the hot well or seal drum. I do this by shutting off the hot well pump and seeing how long it takes the hydrocarbon liquid level to increase by one foot. Then, I sample the hydrocarbon to obtain its molecular weight. Typically, in a refinery vacuum tower, this will be about 140 pounds per mole. Caution – Don't forget about the baffles in the drum.
- Step Ten – For a vacuum tower with no pre-condenser, add in the pounds of steam used in the tower:
  - In the bottom stripping section
  - In side strippers

- Velocity steam or condensate used in the fired heater tube passes

- Step Eleven – For a vacuum tower with a pre-condenser, you must add in the vapor pressure of water at the pre-condenser vapor outlet temperature. For example, the pre-condenser outlet temperature is 102°F. The vapor pressure of water from my steam table at 102°F is 1.0 psia or 51.7 mm of Hg. The observed pressure at the inlet to the first stage jet is 0.2 BAR absolute or 152 mm of Hg.

  Therefore, the moisture content of my vapor is:
  - (51.7) ÷ (152) = 34%

Since your lab sample of seal drum gas was likely reported on a dry basis, this will be a big additional vapor load to the first stage jet.

If you think all this is too complicated, you should have worried about that before you became a process engineer.

Now, you can see if you're on (which is good) the jet performance curve shown in Figure 3.2, or above (which is bad) the curve. Being below the curve is quite improbable. If you are substantially above the curve, the malfunction is likely an ejector mechanical defect or poor motive steam conditions, as I'll discuss in subsequent chapters. Perhaps you're on the curve, but way out on the right. Or worse yet, you're really far out on the right, so that the curve has stopped, meaning, you have overloaded the jet with gas. Now what?

Well, you've got trouble. The non-condensible load to the jet is excessive. The main areas of excess gas are:

- Air leaks in the hot part of the tower.
- Air leaks in the cold part of the tower.
- Excess cracked gas in the feed.
- Excess cracking in tower bottoms.

- Feed contamination with lighter components.
- Heat exchanger leaks.
- Naphthenic acid decomposition.
- Seal leaks on idle pumps.

6.5 Overloading Vacuum Jets

Often, my clients believe that a poor vacuum is due to a mechanical problem with the jets. I had just made that mistake at the Murphy Oil Refinery near my home in New Orleans. I stupidly had them replace one of their second stage jets because I thought the jet was not running on its performance curve. I say stupidly for two really good reasons:

- Reason One – I did not have the manufacturer's curve.
- Reason Two – The new jet didn't help.

As a result of my error, I can now cross Murphy Oil off my list of clients. And that's the way it ought to be.

The first step in checking a jet for overload is to obtain the manufacturer's performance curve. Don't waste your time looking in the project files for this curve. Unlike pump and compressor performance curves, ejector performance curves are not typically supplied with the equipment. You will have to call up the equipment vendor for the curve.

NOTE – The curve only applies at or below the critical or design discharge pressure noted on the curve. If you are above this pressure, you cannot use this curve. The vertical axis of the curve (see Figure 3.2) is the pressure measured at the mixing chamber shown in Figure 2.1. You will need to put a vacuum pressure gauge at this location. If you do not have a pressure connection at this point, but only a half-inch plug, proceed as follows:

- Get a valve with a half-inch screwed end ready.

- With a friend present, loosen the plug to finger-tight.
- Quickly, pull the plug and have your friend screw in the valve. When I say "quickly," I mean one or two seconds.

I would not write this if I hadn't done it myself dozens of times. Needless to say, never ask for management approval for such a procedure.

6.6 Excess Cracked Gas Flow

I had observed in my investigation of a poor vacuum situation that the jet system was overloaded. This overloaded condition indirectly caused the first jet to surge. I say indirectly, because the first stage jet had plenty of spare capacity. It surged only because the final (third) stage jet bogged down (i.e., lost its sonic boost). The problem was that the gas flow to the overloaded third (final) ejector was double the design rate.

6.7 Field Checking Gas Flow Meter in Vacuum Service

My field troubleshooting methods are all intended to fulfill dual objectives:

- Results have to be available fast.
- I have to be able to run the test myself.

For example, at this refinery, the metered flow of gas discharging from the final jet was 28,000 SCF/hour. To verify this number, I pulled out a ¾″ steel plug on the shell-side of the first condenser and replaced it with a ¾″ bushing with a ½″ opening. You may assume that the velocity of air flowing from atmospheric pressure to vacuum condition of less than 300 mm Hg is about 900 to 1,000 feet per second.

Therefore, the volume of air sucked through the bushing is:

$$\frac{(0.20)\cdot 3{,}600\cdot 1{,}000 = 5{,}400\ \text{ft}^3/\text{second}}{144}$$

Where:

- 0.20 is area of bushing's hole, square inches
- 3,600 seconds per hour
- 1,000 is sonic velocity of air
- 144 square inches per square foot

I noted that flow indicator of the off-gas meter jumped by about 10% when I emitted the 5,400 SCF/HR of air. Therefore, the total off-gas flow (i.e., tail gas), must now be about ten times 5,400, or roughly 54,000 SCF/HR. Or, the flow prior to my experiment was about 50,000 SCF/HR. As the metered flow was around 45,000 SCF/HR, I was satisfied that the off-gas flow meter was okay.

And how about emitting air to a hydrocarbon system? Don't worry. At ten percent air (2% O_2), the off-gas is too fuel rich to explode.

At the Murphy Oil Refinery, in order to keep the vacuum from breaking due to off-gas overloading of the ejector system, the console operator would reduce the vacuum heater outlet temperature by 10°F. However, this would allow heavy vacuum gas oil to drop into the vacuum tower bottoms where it could no longer be recovered as valuable virgin gas oil.

I suspected that the cause of the excess off-gas production was too high a liquid level in the vacuum tower bottom's boot. Cracked gas is produced in the liquid phase at the bottom of the boot. The liquid in the boot is around 700°F. The greater the residence time in the hot boot, the greater the rate of off-gas production, due to thermal cracking, that the jets have to handle.

So I placed my vacuum gauge at the bottom of the boot shown in Figure 6.3 at connection "B". It read about 420 mm Hg. The vacuum tower flash zone pressure (see Figure 6.3) was about 40 mm Hg. I calculated that the height of liquid above my gauge in the bottom of the boot was:

$$\frac{(420-40)}{(760)} \times \frac{(34\ \text{ft})}{(0.85\ \text{S.G.})} = 20\ \text{feet}$$

Where:

- The 760 term = atmospheric pressure (760 mm Hg)
- The 34 feet is feet of H20 in one atmosphere
- The 0.85 S.G. is the specific gravity of hot vacuum tower bottoms

The normal boot level should have been about five feet, not twenty feet. Something was really, really wrong.

So I reported to Fred Dow, the operations manager, that the level indicator which was reading 50% or four feet of liquid, was reading very low. The level indicator should have been over-ranged at 100%.

Mr. Dow had the instrument tech recalibrate the level indicator. After the recalibration, the level indicator measured a 40% level. (For a detailed explanation as to how process plant level indicators work, see my book, *Troubleshooting Process Plant Controls*, John Wiley, 2008).

At this point, I went home. My client paid me for one day of consulting services and received nothing for his money.

Who was right? Me or the instrument tech? The answer is we were both right! A year passed during which time my client dropped 700,000 barrels of gas oil into the vacuum

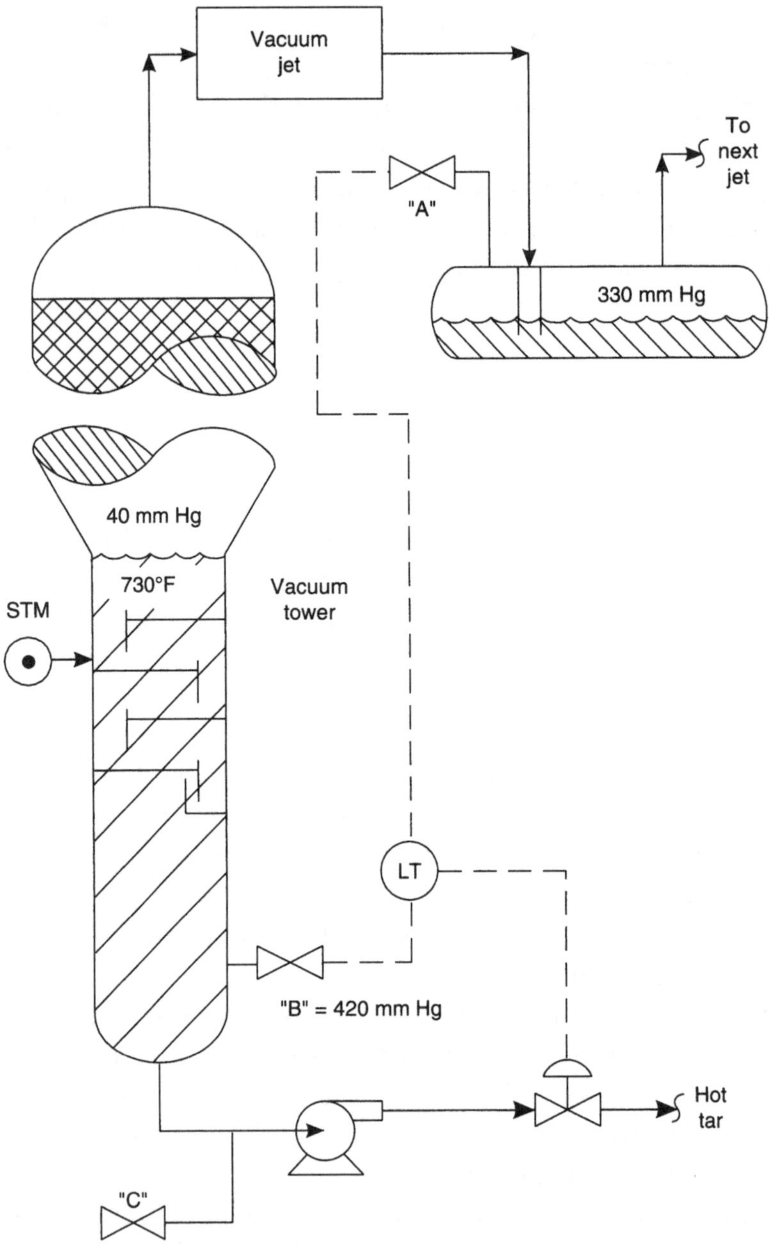

Figure 6.3 Mis-connected bottoms level upper tap causes cracked gas flow to double.

tower's bottom residue product for a net economic loss of \$7 million. That was kind of a lot of money to educate me. But I'm smarter now.

Figure 6.3 explains the situation. The top tap of the level indicator was not connected to the flash zone pressure of 40 mm Hg. Unbelievably, the top tap was connected to a vessel that operated at about 330 mm Hg. So, if I repeat my level calculations shown above, the results are:

$$\frac{(420-330)}{(760)} \times \frac{(34\text{ ft})}{(0.85\text{ S.G.})} = 4\text{ feet}$$

And why did some unknown engineer or instrument tech connect the high pressure (330 mm Hg) to the top of the level transmitter (LT) of the vacuum tower's boot?

"God only knows.

God makes his plans.

The information is unavailable to the mortal man."

Or perhaps it was an insane attempt to keep the top level tap from plugging with tar.

6.8 Surging 3rd Stage Jet Bogs Down Primary Jet

The tiny third stage ejectors did not have very much excess capacity. At first, their soft surging sound was not audible above the mighty roar of the giant first stage jets. But once when I was conducting my air calibration test, described in the preceding section, I did note that the final jets began surging first. Next, the larger second stage jets began surging. Then, one moment later, the first stage jets surged and lost vacuum.

As the final jets (third stage) bogged down due to excessive cracked gas flow, they raised the discharge pressure of the second stage jets above their critical discharge pressure. This caused the second stage jets to also surge.

As the second stage jets bogged down due to excessive back-pressure from the third stage jets, they raised the discharge pressure of the first stage jets above their critical discharge pressure. This caused the first stage jets to surge in turn, and the vacuum tower top vacuum began to break.

Unloading the third stage jets with less cracked off-gas corrected the entire problem at this vacuum tower.

6.9 Exchanger Leaks Overloads Jets

I have two examples of this nasty problem: One, from the now defunct Pacific Refinery north of Berkeley, California. And the second, from the Valero Refinery in Aruba.

In California, the leaking exchanger was a vacuum tower bottoms quench exchanger vs. crude, as shown in Figure 6.4. But, how did I know that the exchanger was leaking crude oil into the circulating quench resid stream?

I had noted that every time the tower bottom's quench flow increased, vacuum was lost. That is, every time Valve "A" (Figure 6.4) opened, the vacuum degraded. One possible explanation was because of increased non-condensable flow to the jets. I therefore reasoned as follows:

- Opening Valve "A" reduced the shell-side pressure in the quench heat exchanger.
- The pressure in the crude side (i.e., the tubes), would then exceed the pressure in the shell. Local pressure gauges supported this hypothesis.
- As the shell pressure fell below the tube pressure, crude charge would leak into the vacuum tower quench oil return stream.
- Light ends in the crude (ethane and propane) would load up the overhead jets and degrade the vacuum.

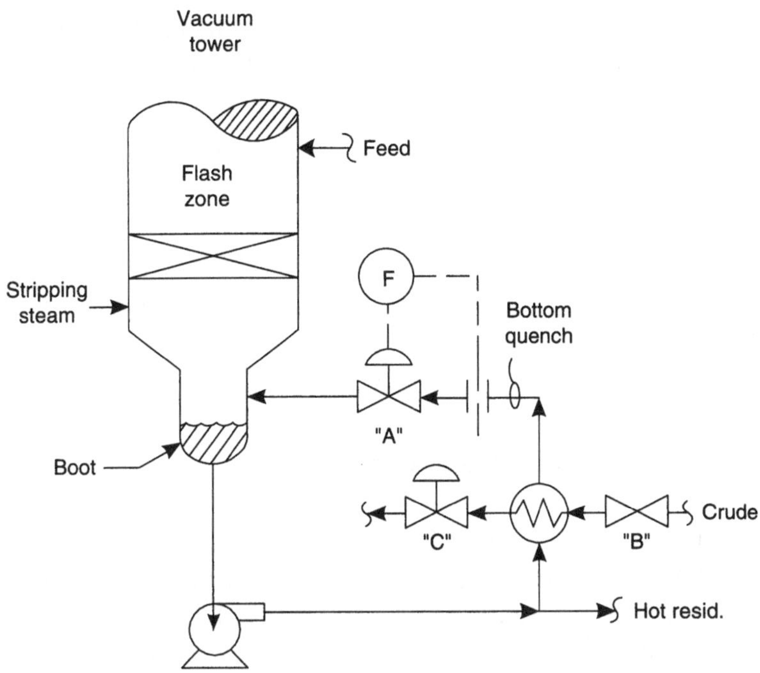

Figure 6.4 Leaking quench exchanger overloads vaccum system with virgin light gas from crude (ethane, propane, butane).

What was my test for my theory?

- Partially close isolation gate Valve "B."
- This caused control Valve "C" to open from 40% to 60% valve position.
- The tube (crude) side pressure in the heat exchanger dropped by 20 psig.
- The tube side pressure was now lower than the shell (quench oil) side pressure.
- The metered vacuum tower off-gas rate dropped, which unloaded the jet system and the vacuum improved.

My story from the Valero Refinery in Aruba, is one of the few occasions when I identified a malfunction by phone. In Aruba, hot vacuum tower feed was used to preheat crude.

The reason for this strange process configuration was that vacuum tower feed flowed through a storage tank operated at 300°F. That's not important. What was important was that the vacuum tower feed vs. crude preheat exchanger was leaking.

But how could I identify this malfunction in Aruba from New Orleans? I asked for a sample analysis (Gas Chromatograph) of the vacuum system off-gas collected from the hot-well or seal drum. This indicated:

- Mole percent propylene = 2%.
- Mole percent propane = 12%.

Normally, the composition of the vacuum tower off-gas reflects the origin of the gas. That is, thermal cracking. In thermally cracked gas, we find that the ratio of olefins to saturates (i.e., propylene to propane), is about 40/60. But in my sample of off-gas from Aruba, the ratio was two to twelve. That is, I had about four times the expected concentration of propane, based on the propylene concentration.

The engineers in Aruba decided to put my leaking heat exchanger theory to a test. They reduced the pressure on the crude side of the suspect exchanger a few psi below the pressure on the vacuum tower feed side of the exchanger. As the vacuum tower feed went through tankage, the vacuum tower pressure, and vacuum off-gas rate slowly declined over the next day. Finally, the off-gas analysis showed that the propane content of the seal drum off-gas had fallen from 12% to 3%. Incidentally, the amounts of ethane and butane (iso and normal) had also declined.

6.10 Poor Vacuum Tower Feed Stripping

I hate Canadians. It's not their fanatical passion for hockey. It's that they live in a cold country and refuse to admit it. For example, I was working at the Suncor Oil Sands Upgrading plant in Ft. McMurray, Alberta, one January. The problem was a poor vacuum on their new, gigantic, vacuum tower. The off-gas analysis showed the following composition:

- Saturated Butanes (Iso plus Normal) = 20 mole percent
- Unsaturated Butanes and Butylenes (C4 Olefins) = 2 mole percent

If the C_4's were just a product of thermal cracking of the vacuum tower feed, the saturate and unsaturated composition would be approximately equal. In this unit, there were no heat exchangers located on the vacuum tower feed. There was no circulating bottom quench, as in the prior example. If the vacuum tower pumparound heat exchangers had been leaking, then the vacuum tower gas oil products would have been contaminated with black tar sands oil (i.e., bitumen). Therefore, I decided that the problem was not a leak of bitumen into the vacuum tower, as in my prior example.

The problem of excess saturated butane in the feed could only be due to poor steam stripping efficiency in the upstream tower. That is, the butanes were not being efficiently stripped-out of the vacuum tower feed. So I increased the stripping steam rate and the malfunction got worse. This meant the pressure in the downstream vacuum tower went up.

Most unfortunately, my increasing the stripping steam rate to the upstream tower also caused this tower's stripper to pressure-up. Why? Because we lacked sufficient overhead condenser capacity to handle the extra steam. And this was on a new unit, using air coolers, in the middle of the winter, with ambient temperatures of minus 20°F!

When I complained that my nose, feet, and fingers were numb from the cold, and when I complained that at minus 20°F, they really should not be limited by air cooler condenser capacity, that's when the unit supervisor, Probkar Reddy, said, "But, Norm, it's been a warm winter. I guess that's why we don't have sufficient stripper condensing capacity to strip the vacuum tower feed properly and prevent the butanes from overloading the first stage vacuum jets."

And that's why I hate Canadians. They live in a minus 20°F environment and claim it's warm. And the worst aspect of the story is that Probkar comes from a subtropical region of India and isn't a native Canadian anyway!

6.11 Level Connection Purges and Pump Mechanical Seal Gas

Often the top tap of level sensing instrumentation is purged with natural gas or nitrogen through a restriction orifice. In my designs, I don't use level control, but pump suction pressure control (as described in my book, *Troubleshooting Process Plant Control*, Wiley Publications, 2008). So for me, it's not a problem. But natural gas is methane, and any purge gas on level taps will contribute to the off-gas handled by the jets. Sometimes pressure sensing connections and flow orifice taps are purged with refinery fuel gas or natural gas. Filling these connections with a light gas oil or diesel oil is sufficient, without gas purging, provided the clean gas oil is changed periodically. I recall that my friend, Dale, who works for Chevron in El Segundo, said they did that perhaps once a month.

In newer pumps in black oil service, we now use nitrogen barrier seals (i.e., dry seals) rather than seal flush oil. I don't really have any idea as to how much N_2 would be used on an idle vacuum tower pump. But, whatever the flow, it will contribute to the off-gas rate to be handled by the ejector system. So, when the pump is on standby, block-in the nitrogen flow to the pump seal. I suppose some clever mechanical engineer will call the seal vendor and tell us the N_2 flow is tiny. But that's only true if the seal face tolerances are as per the manufacturer's specs. And this is a book about malfunctions. So, block-in the N_2 to idle pumps.

6.12 Effect of Heater Outlet Temperature

I remember working in a refinery in West Texas. In the morning it was quite cold. And so is the cooling water to their vacuum tower pre-condenser. As the heat of the day came on, the pre-condenser vapor outlet temperature rose and so did the vapor pressure of water in the pre-condenser vapor flow to the first stage jet. This jet started to surge due to loss of its sonic boost as the vapor load to it escalated due to the extra moles of steam in the pre-condenser vapor effluent. I could hear this surging sound every morning. Also, the vacuum on the tower would increase from about 70 mm of Hg to about 110 mm of Hg. As a result of the degraded tower vacuum, heavy vacuum gas oil production would drop by 2,000 or 3,000 BSD.

To mitigate this problem, I had the operators reduce the heater outlet by 5°F. This relatively small move would usually unload the primary jet enough to restore the sonic boost. If not, and I still heard the surging, I would knock off another 5°F from the heater outlet. The loss of 5°F or 10°F flash zone temperature had a small negative effect on the recovery of heavy gas oil, but was nothing as compared to the 40 mm of Hg drop in vacuum due to the loss of the sonic boost in the first stage ejector.

I'm usually cautious when raising the heater outlet temperature, especially in vacuum service. Certainly, if raising heater outlet by 5°F produces a loss in vacuum of five mm of Hg (0.2″ of Hg), I feel that gas oil recovery will be adversely affected. But even if it's not, and I gain some gas oil production, I would still be concerned about increased coke formation in the heater tubes. The reason I sometimes raise a heater outlet temperature with confidence, is because

I had suppressed coke formation rates with the coil steam. So, regardless of any design criteria or rules of thumb pertaining to the flash zone or heater outlet temperature let me give you some sage advice:

- If the metered off-gas flow increases by 10%–20%, as you increase the heater outlet, back off because you will shorten the heater run length due to coke formation in the tubes.
- More steam in the heater tubes will significantly increase heater run length as limited by coke formation and high tube skin temperatures in the heater tubes.

Incidentally, if you have a pre-condenser, you may be able to reduce the gas load to the first stage by raising the tower top temperature. I noticed this at the Good Hope Refinery

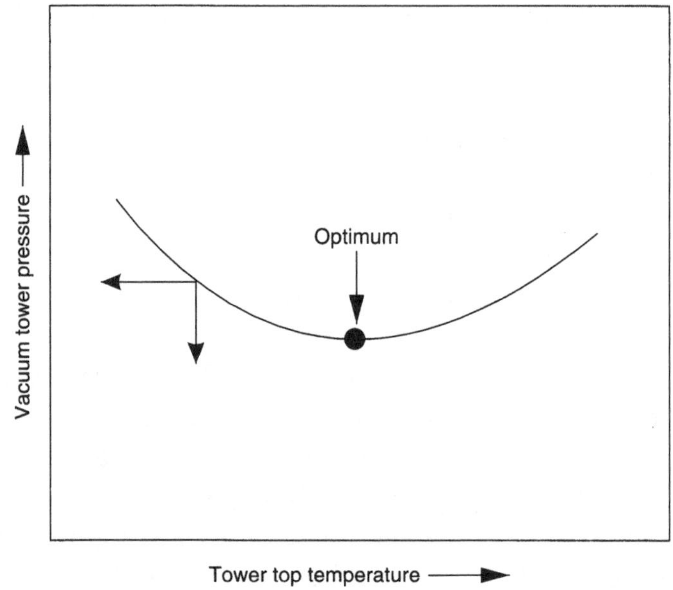

Figure 6.5 For vacuum towers with precondensers, raising the top temperature can unload the jets.

vacuum tower. Raising the top temperature flashes more naphtha overhead, which absorbs some lighter hydrocarbons out of the gas to the first stage jets (see Figure 6.5). Indeed, one of my clients injected some heavy naphtha into the pre-condenser feed for the same purpose. Of course, if the pre-condenser outlet temperature goes up more than a few degrees, you will not gain anything by increased tower top temperature.

6.13 Extracting H_2S from Vacuum Tower Off-Gas Upstream of Ejectors

It was a quiet Saturday afternoon in Barrancabermeja, Colombia, South America. I was lying peacefully on the vinyl couch in my decaying hotel room, watching a giant black roach crawl deliberately across the ceiling. The sunlight filtered softly through the dirty windows. Suddenly, there was a shouting and banging at my door.

"Señor Lieberman! Señor Lieberman! Donde es Señor Lieberman?"

"What's up boys?" I asked.

A half dozen students from my troubleshooting class were crowded around on the rickety hotel balcony.

"Señor Lieberman. You must come quickly. We have made an amazing discovery on our vacuum tower," Freddie, the refinery tech manager, exclaimed.

As we raced to the plant, Freddie said, "Norman, my good friend. Mi amigo. The operator on our vacuum tower has reduced the tower top pressure by ten mm of Hg."

"Como?" I asked.

"He opened wide the neutralizing ammonia to the inlet of the first stage jet," Freddie explained. "It was a miracle. The vacuum tower top pressure dropped by over ten millimeters."

The miracle was short lived. By the time we arrived in the plant, the NH_3 bottle had gone empty. I explained this minor miracle to Freddie and his staff as follows:

- The ammonia flowed through the first stage ejector.
- The cracked gas flow contains 10% to 40% (on a dry basis) of H_2S.
- The H_2S and NH_3 reacted to form water soluble ammonia sulfide NH_4 (HS), which was extracted by the condensing steam in the first stage condenser.
- This unloaded the second stage jet and hence reduced the suction pressure to the second stage jet, which in turn reduced the discharge pressure of the first stage jet.
- At a reduced discharge pressure, the first stage jet picked up its sonic boost and thus drew down the vacuum tower top pressure by ten mm of Hg.

This incident gave me an idea, which I have never tried—to reduce the ejector vapor load by injecting amine into the overhead gas.

I had personally measured the H_2S content of the seal drum off-gas at 30%–40% in Aruba with a dragger tube. Instead of using ammonia, I thought, suppose we scrub-out the H_2S with amine. We will have to scrub-out the H_2S anyway. So, if the second or third stage jets are limiting vacuum, this is a neat way to unload them. The rich amine of course would have to be pumped back to the amine regenerator. One could not expect to reduce the H_2S down to 100 ppm (as normally required) by this technique, but likely, 90%–95% of the H_2S could be extracted. I never convinced the management in Aruba to invest in my idea. If you succeed in doing so at your plant, I would be interested in your results.

6.14 Disposal of Seal Drum Off-Gas

I'm assuming in this section that the off-gas in question is originating in a refinery vacuum tower. A typical off-gas composition (on a dry basis) is:

H_2	5%–10%
N_2	2%–20%
O_2	0%–5%
CO_2	1%–5%
CO	0%–5%
C_1	25%–40%
C_2's	5%–10%
C_3's	5%–15%
C_4's	2%–10%
H_2S	10%–40%

A typical heating valve (dry basis) is 1,200 BTU's per standard cubic foot.

In the U.S.A., it's against the law to burn this off-gas without scrubbing, for H_2S removal although many of my clients do so anyway. Regardless, there are three ways used to compress this gas to a reasonable (5 to 10 psig) pressure, so that it can be run through an amine absorber:

1. A final steam jet.
2. A liquid ring seal compressor.
3. An ejector using higher pressure refinery fuel gas or wet gas.

Method three is bad. One should avoid combining oxygen contaminated gasses with refinery fuel gas. The O_2 will react with the H_2S and form sulfur which will then plug refinery burner tips. The burners will then have to be steamed out to remove residual sulfur deposits. It's best to avoid this problem and do what over half my clients do (at least in the U.S.A.). Scrub the seal drum off-gas in a separate amine tower. Then, burn the off-gas in the vacuum heater via a waste gas burner. Even if your off-gas does not contain much oxygen now, leaks are inevitable.

Note that the 40% H_2S content is consistent with having about 4% sulfur in the vacuum tower bottoms. So, while your off-gas rate will be small, the required amine circulation may be quite large.

This is also a foaming service, and flooding of such absorbers is a common problem, if not designed for about 60% of jet flood and a downcomer loading of about 90 GPM/ft^2.

I attended a meeting last month at which we discussed performance of low pressure waste gas H_2S absorbers. The consensus was that absorption efficiency was poor due to the very low absorber pressures. Next time I have to design such an absorber, I'll design it not for a few psig, but perhaps for 10 or 20 psig. The normal H_2S target is 100 – 150 ppm of H_2S.

6.15 Fouling of Waste Gas Burner

Waste gas burner fouling creates a back pressure on the final stage ejector which may exceed that ejector's critical discharge pressure. This jet can then surge. This raises the discharge pressure of the upstream jet which may also start to surge. Eventually the primary ejector loses its sonic boost and the pressure in the vacuum tower jumps up. The operators say, "The vacuum has broken." It's true the jet is no longer surging, but that's only due to the complete loss of the sonic boost.

My problem is that the surging sound of the small final jets is not very loud, and I don't hear that well. So, you have to be close to these small jets to hear their surging sound. Regardless, a pressure of five or six psig in the seal drum should indicate partial plugging of the waste gas burner. Divert the off-gas to the refinery flare for a brief period (which is okay for emergency maintenance), and clean the waste gas burner.

7

Motive Steam Conditions

When steam expands through the motive steam nozzle, as shown in Figure 1.1, the velocity or kinetic energy of the steam increases. The greater the speed of the steam as it flows into the inlet of the diffuser, which is just a few inches away, the better the job the diffuser can do in compressing the non-condensable vapor. It's the speed of the motive steam that does the compression work, on the moles of non-condensable vapor.

The speed of the steam or the kinetic energy of the steam comes from both:

- The pressure of the steam.
- The enthalpy of the steam.

By enthalpy, I mean heat. Much of the temperature of the steam is converted to the velocity of the steam. If you don't believe me, check the temperature of the mixing chamber of an ejector. It's ambient; you can hold your hand on the body of the un-insulated mixing chamber. The motive steam is typically 150 psig, 400°F superheated steam. Why then, is the steam now only about 90°F? The 310°F (400°F–90°F) or 170 BTU per pound of steam has not been lost. It's been converted into a different form of energy. That is, kinetic energy.

7.1 Effect of Wet Steam

If the motive steam contains moisture, then the moisture will evaporate as it enters the lower pressure in the mixing chamber. For example, if the steam quality is very poor, it might contain 10% water. That amount of water flashing to steam at vacuum conditions would absorb 100 BTU per pound of steam. As the specific heat of steam is 0.55 BTU per pound per °F, the steam would lose:

- 100 BTU ÷ 0.55 = 180°F

The 180°F loss in temperature would mainly be at the expense of the kinetic energy of the motive steam flowing into the diffuser. The slower steam is less able to compress the non-condensable vapors. Thus, the vacuum produced by the jet will be degraded. If the steam gets really moist, the jet will surge. That is, it will lose its sonic boost. As the moisture content of the motive steam to the jet rises, at some point the outlet of the steam nozzle will get so cold, that the droplets of water flowing out of the nozzle will change to ice. The ice will briefly (10–20 seconds) stop the flow of steam.

Don't think I have read this in a book. I have seen steam jets supplied with 350°F steam freeze at a Gulf Coast Refinery (Mobile, Alabama) in August. Also, every time it rained in Norco, Louisiana, I could hear the steam jets surging at the GHR's vacuum tower. Our steam supply lines were poorly insulated and lacked heat tracing.

While superheat is not helpful in improving vacuum, reasonably dry (1–2% moisture) steam is vital. Steam traps do not help very much. The best solution, which I've seen several refineries use, is to have a knock-out drum on the steam supply, equipped with a demister, located in close proximity to the steam jets.

7.2 Water in Motive Steam

I'm sitting on the beautiful beach in St. Croix in the Virgin Islands. I've completed a difficult week working for the Plant, the world's 8th largest refinery, located across the island, and visible from my hotel.

On one tower, I improved the vacuum top pressure from 13 mm Hg to 8 mm Hg. I had observed moisture condensing on the converging section of the primary ejector, as shown in Figure 7.1. In St. Croix, the humidity is typically 100%. Ambient temperatures are 90°F. Moisture (and in cooler climates, ice), forming on the outside of the converging section of the jet, is a positive indication of extremely wet steam (or liquids in the

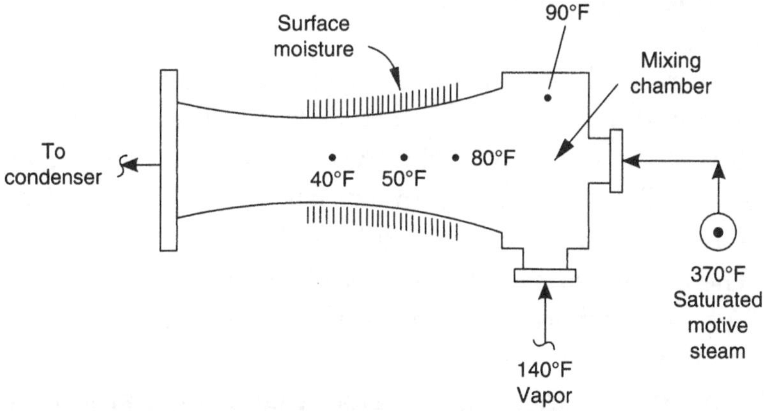

Figure 7.1 Moisture condensing on the outside of a jet is a sure sign of wet steam. The temperatures shown are all skin temperatures.

feed). The moisture in the motive steam evaporates in the low pressure zone at the inlet to the converging section of the jet. Based on the observed skin temperatures shown in Figure 7.1, I concluded that perhaps 5–10 wt.% of the steam was water. The evaporation of the water extracted a lot of heat from the motive steam.

The odd thing about this situation was that the motive steam was flowing from a steam source that was superheated to 500°F. This is 130°F above the steam's saturation temperature. However, I then discovered that a De-Superheating Station had been installed on the steam supply line to the ejectors. In theory, excessive superheat in steam degrades the performance of a steam jet. In practice, the effect of superheat appears small.

However, my client, thinking that avoiding superheat in steam was critical, had an excessive amount of water injected into the motive steam. I then had the de-superheating water shut off. The pressure at the top of the vacuum tower dropped by 5 mm Hg.

7.3 The Tale of Weak Steam

Sometimes I get paid quite a lot for very little. Here's such an example.

Quite recently I had a project to improve the vacuum on a column at a refinery in Corpus Christi, Texas. I climbed to the top of the tower with the two operators. I could hear the three parallel first stage primary jets surging before I could see them. The vacuum was really bad.

"The jets are surging," I observed the obvious.

"Yes, Mr. Lieberman, they do that a lot," said Tommy, the younger operator.

"What do you mean 'a lot'? Does that mean that sometimes they don't surge?" I asked.

"Well, Mr. Lieberman," answered J.J., the older operator. "Not when we have strong steam. Then the jets don't surge. We get two types of steam here. Weak steam and strong steam. It's the weak steam that causes the jets to surge and hurts vacuum."

"You mean the weak steam is at a lower pressure than the strong steam?" I tried to clarify.

"No, No!" Tommy shouted above the jets. "It's all the same 400 psig steam. We get the weak 400-pound steam from our No. 5 boiler. We get the strong 400 pound steam from the Union Carbide plant across Navigation Boulevard. It's all the same steam. It's just that the Carbide steam is strong and our steam is weak. All the steam we get is at the same temperature and pressure. It's like me and J.J. He's weak, and I'm strong, even though we're both married," Tommy laughed.

The strong steam was dry, saturated steam. The weak steam had a very high moisture content. I recommended to my client that they reinstall the inline steam water separator that had been removed many years previously. And for this they paid me $5,000 (less the $20 for pizza I bought for Tommy and J.J.).

7.4 Internal Freezing of Steam Nozzle

I just got off the phone with my pal, Scott Golden. Scott and I had an argument.

Scott said, "Norm. Are you crazy? You can't write a book about troubleshooting vacuum systems. There are 1,000 different types of malfunctions."

And I said, "Scott. Man is the greatest factor in the Universe. He can do anything. Anyway, there are only five hundred different types of malfunctions."

But the more I reflect on vacuum system malfunctions, the more I'm beginning to think that Scott Golden is right. For example, Liz and I were working on an asphalt vacuum tower along the Chickasaw River in Mobile, Alabama.

The primary ejector was making a surging sound, indicating a periodic loss of its sonic boost. However, I also noted the following:

- The jet's discharge pressure was well below its design critical discharge pressure.
- My client had recently replaced the jet's steam nozzle, even though the old one was in perfect condition.
- The gas rate to the jet was half of its design value.
- The motive steam pressure was at design.

At each surge, the pressure in the vacuum tower jumped far more than I expected from the loss of the sonic boost. Also, changing the jet's discharge pressure or its gas rate did not alter the two minute frequency of the surges.

Liz, who was monitoring the skin temperature of the jet's mixing chamber (see Figure 7.1), noted the following:

- The vacuum would break every two minutes as the jet went very, very quiet.
- The mixing chamber temperature would drop from about 70°F to exactly 32°F.
- After remaining quiet and cold for a minute, the jet would suddenly come to life with a roaring sound for a minute. It would heat back up and the vacuum would be briefly restored.
- The cycle would be repeated with great regularity every two minutes no matter what other parameters we varied.

When the front portion of the ejector chilled to precisely 32°F (0°C), it did not become just quiet, it became silent. This meant that the motive steam flow through the inlet nozzle had stopped. It had frozen solid with ice.

Normally the motive steam cools to about 90°F to 100°F downstream of the steam nozzle. But if there is about 5%–10% entrained water in the motive steam, then the expanding steam can form ice and plug the inlet nozzle. The ejector stops working completely. Both the sonic boost and the velocity boost are completely lost. The warm vapor in the downstream condenser now flows backwards through the ejector. This begins to break the vacuum in the tower, but also melts the ice in the steam nozzle. The motive steam flow to the jet is thus restored. The jet starts to work again, and the cycle is repeated.

How can I be so sure the problem was excessively wet steam? Well, because the unit was then shut down. A steam strainer device (which removes the bulk of the water from steam) was installed. After restarting the vacuum tower, the erratic vacuum was no longer observed.

And the reason I think Scott Golden is probably right about the 1,000 malfunctions, is that I'm only 70, and I've only likely seen half the problems so far. If I can keep on going until I reach 100, I'll probably observe Scott's 1,000 vacuum system malfunctions.

A final note about the Chicksaw River incident: Several years later, I was told that the steam inlet nozzle was of a multi-port design. Meaning instead of perhaps a single one inch opening, it might have had sixteen ¼″ openings. The smaller ports could have plugged far more easily than a single larger port.

I never saw the ejector's internals myself, and have never heard before or since, about such a multi-port steam nozzle. It seems like a really bad design feature, not only because of ice, but because of potential hardness deposits (i.e., silicates) in the motive steam supply to the ejectors.

7.5 High Pressure, Superheated Motive Steam

Here's a word of caution about steam conditions: Often, reducing the motive steam pressure below the design pressure improves vacuum. This needs to be determined by trial and error. Of course, if you get the steam pressure too low, you will cause the jet to surge and break the vacuum.

On the other hand, I've never seen vacuum improved very much by exceeding the jet name plate motive steam pressure. Usually, excessive motive steam pressure causes a loss in vacuum.

This is the same for superheated steam. While dry steam is critical, as I've discussed in this chapter, superheated steam does not help. If you have higher pressure, superheated steam, you can modify (or replace) your existing jets for the new steam conditions. This will result in a reduction of the pounds of motive steam consumed for each pound of non-condensable gas compressed in the ejector system.

7.6 Effect of Moisture Content of Saturated Steam on Temperature

I've been teaching a course for a Tar Sands Upgrader Facility in Alberta this week. Tommy, one of the participants, asked the following question:

"Norm. Our ten bar steam is typically superheated by 10°C–20°C. But sometimes it cools to 190°C, and becomes saturated. Sometimes the saturated steam doesn't make much difference, but sometimes it really hurts the vacuum a lot."

And I asked, "What do you think happens to the temperature of saturated steam when half of it condenses?"

And Tommy answered, "I guess it cools off a lot because it's partly condensed. Is that right, Norm?"

I have had over 17,000 engineers and operators attend my process troubleshooting seminars. About ⅔ of the operators and at least ⅓ of the engineers do not know that saturated steam at a given pressure has the exact same temperature regardless of its water content because it is a pure component. Certainly, Tommy's problem was the variable amount of water in his steam supply to his ejectors, not the temperature.

I guess it's a lesson that I've learned before. Don't assume either engineers or operators understand basic physical concepts. When you ASS-U-ME, you make an "ASS" out of "U" and "ME."

7.7 Steam Pressure Affects Vacuum

The steam jet manufacturer can design jets to function over a wide range of steam pressures. This is not to say that he can design a jet to work equally well with varying steam pressures. The jet must be designed to work at a particular steam pressure.

On one unit, the designer knew that the refinery steam pressure would vary between 130 psig and 170 psig. He decided to design the steam jets for 150 psig steam. Aware that the jets' performance would suffer at lower steam pressures, he added a spare jet in parallel with the basic steam jets. His idea was to turn on the spare jet whenever the refinery steam pressure dropped below 150 psig.

Unfortunately, steam jets do not work this way. A jet that is designed for 150 psig will gradually draw less vacuum as the steam pressure falls to 140 psig. Below 140 psig, the performance will drop precipitously. On this particular unit, here is what happened:

Steam Pressure (psig)	Vacuum (mm of Hg)
160	100
150	100
140	125
130	350

The correct way to handle the problem of variable steam pressure to a vacuum system is shown in Figure 7.2. The steam pressure is maintained by throttling the steam through a control valve. A jet designed to use 130 psig steam will consume about 15% more steam than a jet designed to use 150 psig steam.

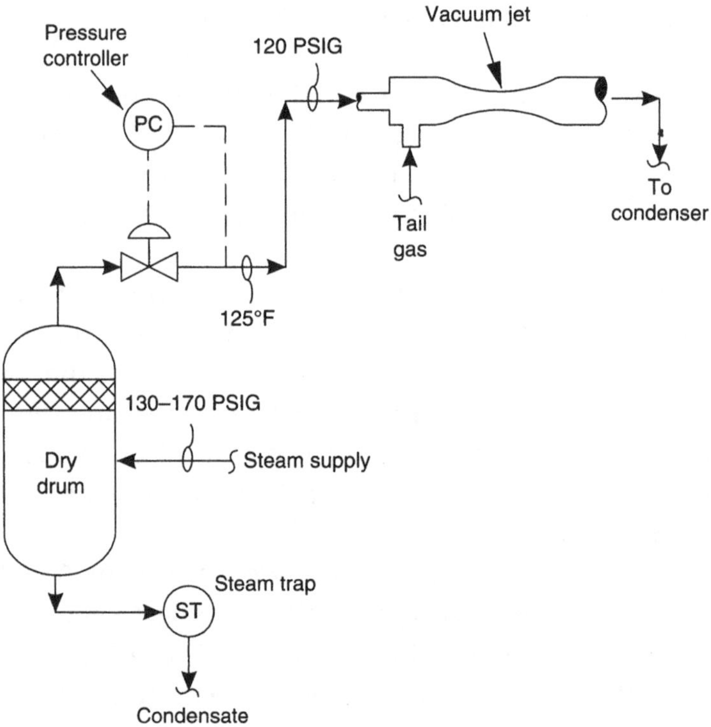

Figure 7.2 Wet steam of variable pressure will ruin a vacuum steam jet's performance.

Why not simply design the jet for 130 psig, save the cost of the control valve shown in Figure 10.2, and reap the benefits of deeper vacuum whenever the steam pressure is higher than design? This approach is false because:

1. A steam jet works best with the design steam pressure.
2. Higher steam pressure forces more steam through the jet; however the incremental steam flow is wasted and can reduce the vacuum.

7.8 Effect of Superheated Steam

Superheated steam does contain more energy than saturated steam. Thus you would think it could produce a better vacuum.

However, the specific volume of superheated steam is greater than saturated steam. Thus, a steam nozzle of a given size would pass fewer pounds of steam for superheated vs. saturated steam.

The benefits of superheated steam are thus more than outweighed by the greater density of saturated steam. In practice, the effect is small compared to other problems I've observed.

I've quantified the above in Chapter 17, "Frequently Asked Questions".

8

Mechanical Defects of Ejectors

I've described in the previous chapter how wet steam can cause the immediate loss of the sonic boost and jet surging. Or, as the operators say, the vacuum has broken. But wet steam also has another, longer term detrimental effect on the ejector. That is, erosion of the interior of the steam nozzle due to extremely high velocity droplets of moisture impinging on the surface of the nozzle. The confusing aspect of this problem is that the nozzle does not appear, when visually inspected, to be damaged. Apparently, the droplets of water erode the nozzle interior in a very even and smooth fashion. The relatively small increase in nozzle diameter results in the loss of jet performance.

8.1 Steam Nozzle Testing

How then, can one determine before taking a jet out of service and disassembling the nozzle, that the problem is uniquely due to an eroded motive steam nozzle?

- Step One – Place a pressure gauge on the inlet to the downstream condenser.
- Step Two – Place a pressure gauge on the motive steam inlet.
- Step Three – Place a pressure gauge on the non-condensable inlet nozzle.
- Step Four – Slowly close the steam inlet valve. Reduce the motive steam pressure to perhaps 70% to 80% of the design motive steam pressure. This pressure is stamped on the body of the jet.
- Step Five – If the pressure at the non-condensable inlet gets lower (i.e., the vacuum is getting better), then the motive steam nozzle is eroded and needs to be replaced.
- Step Six – Don't jump to conclusions. It could be the vacuum has improved because the jet discharge pressure has dropped. Check the condenser inlet pressure. If this pressure has also decreased, then you have unloaded the condenser due to decreased steam load. Likely, this may indicate that this condenser is fouled or undersized for the service.

This above procedure is also used to optimize the steam flow to jets, regardless of their current physical condition. The optimum operating flowing steam pressure is not necessarily the design motive steam pressure. One has to experiment.

8.2 Other Mechanical Defects of Jets

Sometimes, all one can say about a jet malfunction, is that the problem is of an unknown mechanical nature and the jet has to be taken off-line and inspected. For example:

- In one refinery, the diffuser was installed in a reverse position. That is, the diverging section was upstream of the converging section. I had discovered that the motive steam was flowing backwards, out of the inlet nozzle of this jet and then into the inlet of a parallel jet. Isolating this misconnected jet (closing the process inlet and the motive steam valves) had vastly improved vacuum. What the actual mechanical malfunction was, I had no idea at the time. I was informed just recently by a student in my seminar about the reversed position of the ejector. The incident became famous in her refinery, even though it all happened over a decade ago (see Figure 8.1).
- In the same refinery, at the same time, I blocked in another one of the three parallel jets, basically for the same reason. The observed jet "inlet" nozzle temperature was hotter than the upstream

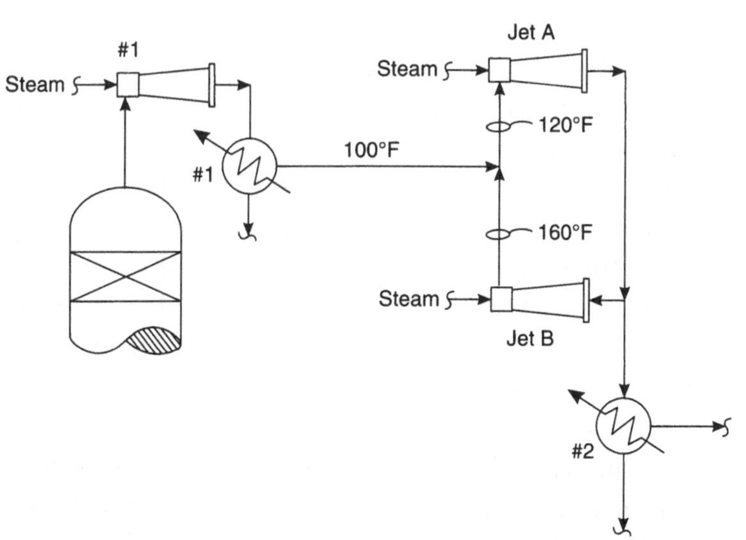

Figure 8.1 A malfunctioning secondary jet. (steam to jet B is flowing backwards.)

vacuum tower top temperature. This is an obvious indication of reverse steam flow through the jet. However, for this jet, the mechanical malfunction was of a far more ordinary nature. Hardness deposits from poor quality steam had accumulated in the steam nozzle. This is quite an easy malfunction to identify and correct. The steam nozzle (which is extremely small as compared to the whole ejector) can be unbolted or unscrewed from the back of the ejector. Some jets are equipped with a clean-out plug in the back of the motive steam nozzle for removing such hardness deposits without removal of the steam nozzle. The hardness deposits are not inevitable. They are a consequence of poor level control in waste heat kettle boilers, and other plant steam generation equipment that permits dissolved salts in boiler feed water (i.e., silicates) to be entrained into the produced steam.

- At a refinery in Denver, I spent the longest time trying to define the problem with a primary jet on a small vacuum tower. Finally, I conceded defeat and informed my client that the jet was suffering from a mechanical malfunction. It just refused to operate anywhere near its performance curve regardless of motive steam conditions, non-condensable vapor load, discharge pressure, or the phases of the moon. Even prayer had proved ineffective. After I left, the vacuum tower was shut down and the jet disassembled. My client informed me that the spacing between the end of the steam nozzle (inside the mixing chamber) and the inlet to the diffuser was adjustable. I had never heard of such an adjustment before. However, I do know that this dimension is critical for proper jet performance. My client, with the manufacturer's representative guidance, corrected the mal-adjustment and then the jet worked just fine.

- At the Good Hope Refinery, where I was Tech Manager, we found that a vacuum jet on our visbreaker residue vacuum flasher was not working. The maintenance crew disassembled the jet for inspection. They found the problem. The threads, where the steam nozzle was screwed into the back on the mixing chamber, were badly eroded. This allowed a portion of the motive steam to bypass the steam nozzle and flow directly into the mixing chamber, shown in Figure 2.1. The problem was that we didn't have a spare nozzle, and a new nozzle was thousands of dollars and many weeks away. So I, your heroic author, wrapped half a roll of Teflon tape around the threads. The steam nozzle was forcefully screwed back into the mixing chamber and the jet was successfully returned to service. And if you visit my home in New Orleans, you'll find that I've employed this sort of advanced repair technology at many locations in my house.

8.3 Fouled Steam Nozzles

I received a nice letter via email last year from a client.

Hi Norman:

I want to share a success with you. I was working on the ejector system of surface condensers in one of our plants. Through various testing, I was convinced it was the ejector nozzles that were a problem. I found nozzles fouled with carbonate deposits.

Lots of gunk, which I filed out and returned the nozzles to service. Saving about 90 T/H superheated steam = 20 MW electricity! (Author's Note – I later found that eight systems were involved and this huge electric savings was a sum of improving all of these eight systems.)

Warwick Hayes
Process Engineer - 5050/
South Africa

8.4 Diffuser Erosion

While erosion to the motive steam nozzle is a pervasive problem, erosion to the downstream diffuser is also common. For example, on our primary steam jets at the Good Hope Refinery, the neutralizing amine had been injected in the wrong place. The jets were placed on top of the vacuum tower without an intervening pre-condenser. The non-condensables from refinery vacuum towers contain HCl (hydrochloric acid). It can't be avoided. The HCl is formed from the hydrolysis of $MgCl_2$ and $CaCl_2$ in the atmospheric tower bottoms. That's why we should inject the neutralizing chemical or NH_3 into the overhead vapor line. But in this case, the neutralizing chemical was injected downstream of the diffuser. A good place to introduce the small neutralizing flow would be into the motive steam. That's where I switched it to, after we eroded a hole into the diffuser body. The hole sucked air into the diffuser and caused the jet to surge.

The interesting aspect of this problem is that all we did was fix the hole to restore the ejector so that it operated on the vendor performance curve. We wrapped it with duct tape. I imagine the interior of the whole diffuser must have been badly eroded by the wet HCl attack. But, this did not appear to degrade its performance. On the other hand, almost unnoticeable amounts of wear on the steam nozzle, severely affects the performance of an ejector.

Incidentally, I said I switched the injection point for the neutralizing amine. Currently, I would substitute NH_3 for the amine. While more difficult to handle, ammonia is a fraction of the cost of the very expensive neutralizing amines that are sold to us by our friendly, local, chemical vendor.

The above story is also a strong argument for double-desalting on the crude unit. The HCl evolved will be reduced by an order of magnitude.

8.5 Repair of Ejector Body

Erosion of an ejector steam nozzle requires the nozzle to be replaced. Erosion of the diverging-converging section (i.e., the main mechanical component of the ejector) will not affect performance very much, until a hole is worn through the ejector's body. The correct way to effect the repair is to wrap duct tape around the thinned area. This is a lot cheaper and quicker than sending the ejector body off to the factory for repair. I've seen this done at the Calumet Specialty Refinery in Shreveport, Louisiana, and it worked just fine – at least for a year or so.

The converging-diverging section of an ejector probably represents 95% plus of the weight of an ejector assembly. The steam nozzle, which is the critical portion of the assembly, is probably less than 5% of the ejector's weight. Removal of the steam nozzle from the rest of the assembly is very easily effected.

8.6 Changing Worn Steam Nozzles

There is a large refinery in Corpus Christi with a vacuum tower overhead system that normally runs a very deep

5–7 mm Hg. During the winter months, the vacuum system would run smooth and steady at the deep vacuum. The summer months would always bring a constant struggle to keep the primary jets in the "boost" operation. In the spring, the jets would swing in and out of the sonic boost from day to night. In the summer, the system would lose its sonic boost for months, and then swing back in the fall.

I observed that:

- The discharge pressure of the primary jet was at the design discharge pressure.
- The pressure drop across the primary condenser was at design of 5 mm Hg.
- One of the three small parallel 3rd stage ejectors was flowing backwards and could not be blocked in.
- One of the three 2nd stage ejectors was the first ejector to lose its sonic boost when the system went unstable and out of the sonic boost operation.
- All steam and cooling water conditions were within design range.

The two bad ejector nozzles were both pulled. Looking very closely, both of the nozzles had very small grooving and wear. Upon installation of new nozzles and start-up, the unit ran continuously in the sonic boost operation for several years, with the primary stage discharge pressure maintained comfortably 5–10 mm Hg below the design of 75 mm Hg, and the vacuum tower top pressure maintaining a steady 6 mm Hg.

8.7 Restoring Critical Flow

Sometimes, two parallel jets do not work as a team. Jet A in Figure 8.1 was a really strong worker. Jet B was a loafer. It is rather like running two centrifugal pumps in parallel. Unless both pumps can develop about the same feet of head, the strong pump takes all the flow, and the weak pump is damaged by internal recirculation. In the case of jets working in parallel, the strong jet takes all the gas flow from the upstream condenser. Furthermore, the strong jet sucks motive steam out of the mixing chamber (see Figure 1.1). As you can see in Figure 8.1, the suction temperature of jet A is 20°F hotter than the condenser number one vapor outlet temperature. This could happen only if the flow of vapor in the jet B suction line was backward. Blocking in jet B stopped the primary jet from surging, and the vacuum tower performance vastly improved.

Regardless of the cause of the loss of the sonic boost, one can identify easily, that the sonic boost is lost. Check the temperature increase from the mixing chamber to the diffuser throat. If the temperature rise is 50°F to 100°F, then the sonic boost is still present. If this temperature rise is 5°F to 10°F, then the sonic boost has been lost. Regardless, the temperature increase between the diffuser throat and the jet discharge will still be another 50°F to 80°F due to the velocity boost.

9

Condenser Fouling and Cleaning

Fouling on the shell-side on surface condensers is not a big problem in steam turbine surface condensers. However, it is a giant problem in refinery vacuum tower condensers, especially the first and last stage condensers. Such condensers have to be mechanically cleaned every few years by pulling the tube bundle.

9.1 Fouling Mechanism in Condensers for Refinery Vacuum Towers

A study by one of my clients indicated that refinery vacuum tower overhead condenser fouling is due to:

- Thermal cracking in the vacuum heater produced di-olefins (i.e., hydrocarbons with two double bonds).

- The di-olefins reacted with oxygen introduced by air leaks in the vacuum tower to produce polymers (i.e., gums).
- The gums caused particulate matter to adhere to the cooling water tube exterior surfaces.

Lab experimentation found that the resulting black, organic deposits were very soluble in methanol, but far less soluble in heavier alcohols. The problem with methanol is that it will vaporize at normal operating conditions. Thus, it can only be used as an off-line cleaning agent.

9.2 Fouling Due to Chemical Additives

Corrosion control chemicals, such as filming amines, may be incompatible with the vacuum tower overhead vapors. Condenser shell-side fouling may result. The deposits look somewhat like coke, but of course, no coke can deposit in these sections of the vacuum overhead system.

The black material will be quite insoluble in water. Lab analysis will indicate a relatively large concentration of nitrogen, which is a typical degradation product of many corrosion control chemicals. The analysis may also indicate ten or twenty thousand ppm of iron, and lesser amounts of nickel, chrome, and moly. These metals are simply products of corrosion and are not the source of the fouling problem.

I do not know of any effective methods to remove these corrosion control chemical fouling deposits on-stream. At Texas City, I found that steaming out the affected equipment would melt the deposits for the particular chemical involved. I recall that the inhibitor chemical was sold by Chemlink (1975), so I suppose I should forget and forgive the nasty experience after 37 years have passed.

9.3 Minimizing Condenser Fouling in Vacuum Towers

The following steps should directionally reduce the rate of shell-side condenser fouling:

1. Minimize the tower top temperature. This should help suppress the carry-over of the heavier di-olefinic compounds from the light vacuum gas oil pumparound section. However, too low a temperature (below water dew-point) will cause corrosion in the vacuum tower top pumparound.
2. Avoid excessive (over 40–50 psi) delta P through the top pumparound return spray header. Higher pressure drops promote mist formation from the spray nozzles and entrainment of the heavier di-olefin compounds.
3. Reduce the heater outlet temperature. Cracking and formation of di-olefins increase exponentially with temperature. Use of steam in the vacuum heater passes can be used to offset the loss of gas oil lift due to the lower flash zone temperature.
4. Air leaks are the root cause of much of the fouling problems, as they convert the di-olefins into polymers (i.e., the black, sticky, greasy, viscous material found in condensers).
5. Try using NH_3 rather than the more expensive neutralizing amines, which should be less subject to promoting nitrogen fouling deposits. Also, be cautious about the use of filming amines, which I often suspect contribute a lot to fouling.

When I wrote above that condensers were a serious fouling service, this statement has to be viewed in perspective. I'm currently working with BP on a terrifically fouled condenser, which has not been cleaned after five years of service. If they had designed the condenser for online yearly cleaning, then their fouling problem would not be viewed as all that serious. But they cannot clean the shell-side of the condenser at all. They have used a fixed-tubesheet design and the tube bundle can never, ever, be extracted from its shell!

9.4 Fouled Pre-condenser

No. 11 Pipe Still at the Epcot Refinery in Baton Rouge produced 12,000 BSD of paving grade asphalt. However, as the unit supervisor, Jerry, explained, "Norman, due to a bad vacuum, we have been unable to extract the lighter gas oil components from our asphalt product. During our last turnaround, Kevin O'Brien, my predecessor as unit supervisor, added a parallel set of steam jets to improve the vacuum."

"That was a stupid thing to do, Jerry," I said. "Look at the pre-condenser vapor outlet temperature and pressure (Table 4.1 and Figure 9.1) that we've just measured. You're limited by the 51 mm of Hg vapor pressure of water at the pre-condenser 100°F vapor outlet temperature."

"Which means?" asked Jerry.

"Which means that the only way to reduce the pre-condenser operating pressure is to lower the condensing pressure of steam inside the condenser by having the steam condense at a lower temperature."

"I guess, Norman, that's why when I commissioned the new set of jets last weekend, it didn't improve the vacuum at all," Jerry said.

"Jerry, more ejector capacity downstream of the pre-condenser can't help you."

"But, Norman," Jerry still objected, "Epcot has produced paving asphalt at No. 11 Pipe Still for years. How come I can't do it now? The pre-condenser is the same."

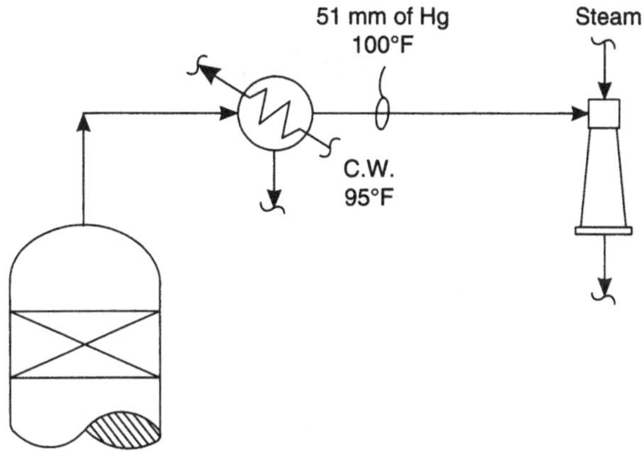

Figure 9.1 Asphalt tower limited by the vapor pressure of water. Improving the jet performance will not improve the vacuum.

"Progressive shell-side fouling," I answered.

"But just last January, when Kevin had my job, he was still able to meet asphalt spec's," Jerry said.

"The cooling water was a lot colder in January than it is now in the middle of August. Jerry. The Mississippi River was 44°F off Algier's Point five months ago," I explained.

"So, Norman. My problem is a fouled pre-condenser. Right? I'll shut down and hydroblast the water side of the condenser's tubes."

"Nope. Won't help. Kevin had the tubes inspected before. They were clean, but he hydroblasted them anyway. Didn't help."

"Okay. So the fouling must be on the shell-side. Is that common, Norman?"

"Oh yeah. The fouling is on the shell-side. Polymers, corrosion products, heavy green waxy stuff. That's for sure. It's on the shell-side Jerry."

"So I'll pull the bundles and have the shell-side cleaned. That's no big deal."

"No Jerry. This is a fixed tubesheet design. The bundle is permanently fixed inside the shell. Can't be pulled."

"What am I going to do until next winter when the river gets cold again? I've got to make paving grade asphalt this summer."

"The only quick fix is to chemically clean the shell-side in place. Let me refer you to the experience at the Citgo Refinery in Lake Charles, who also had a shell-side fouling problem in a vacuum overhead condenser. Maybe you can draw from their experience. However, the real answer is to use an AES type exchanger (i.e., a pull-through floating head bundle) and never a fixed tubesheet design. Use one inch tubes, rotated square pitch, on 1½ inch center to promote ease of cleaning," I said.

9.5 Fixed Tube Sheet Condensers

For condensers in refinery vacuum tower overhead service, the ability to clean the shell-side is critical. Both for the pre-condenser and jet interstage exchangers, fouling is certain. The cracked olefinic gas contains di-olefins. In the presence of oxygen from inevitable leaks, free radicals will form. These free radicals will polymerize to form gums. The gums will mix with corrosion products produced from HCl and H_2S, and with the sulfur compounds produced from the partial oxidation of H_2S. This results in a sticky, black, somewhat waxy mess on the shell-side. I have not had any particular success in chemically cleaning such deposits.

Fixed tubesheet design exchangers are often offered by the ejector vendors as a less expensive alternate to an ordinary pull through bundle with a floating head (TEMA designation "AES"). Such exchangers can only be cleaned on the tube side, which is fine for a surface condenser handling a silicate free steam turbine exhaust. A fixed tubesheet exchanger cannot be cleaned on the shell-side for the simple reason that the bundle cannot be extracted from the shell.

If you ignore this advice, you will eventually replace the fixed tubesheet exchanger anyway, with an exchanger where the bundle can be pulled for cleaning. All my clients, sooner or later, come to this unpleasant realization. So, you might as well save some money and lost gas oil recovery due to poor vacuum and order the correct sort of condensers in the first place. Just specify "AXS" on your bid specs.

Incidentally, the condenser that is most prone to fouling on the shell-side, is the final condenser on the discharge of the last stage jet. Why this is so, I cannot say, but the problem is almost universal.

9.6 Cleaning Condensers On-Stream

I have never actually cleaned a vacuum tower surface condenser while it was in service, and suspect this may not be feasible for many locations. But, what can be done on-stream is outlined below:

- Step One – Install a bypass around the condenser.
- Step Two – Shut off the motive steam to the upstream jet.
- Step Three – Block in the condenser. Blinding is not necessary.
- Step Four – Assisted by a small steam flow, inject a chemical that will dissolve greasy, hydrocarbon deposits. Cool methanol is another alternate.

In a typical three stage jet system, this will remove one of the three stages from service. Considering that the impaired effectiveness of the temporarily disabled stage was reduced by the fouled downstream condenser, and the high delta P of this condenser, the affect may not be unacceptable for one of two shifts. In one case, where a first stage inter-condenser was bypassed and cleaned, the flash zone pressure almost doubled during the cleaning. However, with two years to go to the next scheduled unit turnaround, this was an acceptable short-term loss to the refinery.

9.7 Optimum Condenser Bundle Configuration

The anticipated effectiveness of such cleaning attempts is a function of the shell-side bundle configuration. One inch (O.D.) tubes, set on a rotated square, 1½" pitch is most amenable to cleaning. Three quarter inch O.D. tubes that are set on a triangular 15/16" tube pitch, will be far more difficult to clean in a reasonable amount of time. Obviously, the larger the diameter of the bundle, the more difficult cleaning the latter bundle configuration becomes. One of my clients on the Gulf Coast had a 72" I.D. shell with the above 15/16" triangular pitch. Repeated attempts to clean this bundle (which was also of a fixed tubesheet configuration), were largely unsuccessful. Ultimately, the condenser was entirely replaced with a pull-through, floating head (i.e., TEMA AXS) type exchanger.

9.8 Chemically Cleaning Condensers (Contributed by Gerry Obluda)

The delta P across a primary condenser was 25 mm of Hg, three times above design. The primary ejector discharge pressure was bumping against the design pressure. When the system would suddenly lose vacuum, there was no indication from anywhere other than one of the three primary ejectors breaking.

During a recent turnaround, the ejector nozzles had been replaced and it was unlikely that mechanical damage was to blame. The refinery had diligently maintained steam and cooling water quality, but had been unable to pull the large primary condensers for shell-side cleaning during the recent turnaround due to time constraints.

The condenser delta P had risen from the design of 10 mm Hg to 25 mm Hg. This change alone had driven the primary jet discharge pressure from 65 mm Hg to 80 mm Hg with a design discharge of 75 mm Hg. The primary ejectors could not operate at a higher than design discharge pressure and maintain the sonic boost operation.

Refined Technologies, Inc. (www.r-t-i.com), that provides specialty solvent chemicals for vapor phase cleaning of exchangers, was retained to clean the condensers. During the short outage, the refinery was able to continue to run steam through the primary ejectors and through the shell-side of the condensers. They injected the RTI chemical with the steam, cut back on the water to the primary condensers to ensure high vapor flow rate throughout the exchanger, and ran in that condition for several hours.

Upon start-up, the primary condensers' DP was restored to design, allowing the primary ejectors to discharge back below their design point. The system now ran steadily in the sonic boost mode.

9.9 Ball Cleaning Condenser Tubes (Contributed by an Operator at the Syncrude Plant in Canada)

Ball cleaning skid is on automatic control, and circulates balls for 180 minutes. At the end of 180 minutes, the discharge valve from the collection pot will close. This causes the balls to collect. Balls are collected for 120 minutes. At the end of 120 minutes, the pump shuts down, the collection screen closes, and the skid sits idle for 240 minutes to reduce fouling of the screen.

This happens while the unit is on-line, with enough flow and velocity to maintain circulation.

For information on the balls, visit http://www.toprogge.de.

Before installing this system, it was necessary to acid or mechanically clean the exchanger every two weeks due to rapid fouling. Since installing the Ball Cleaner in 2005, only one mechanical cleaning has been required.

9.10 Corrosion Control by Better Desalting

Corrosion and consequent fouling of a crude unit vacuum system is best controlled, not with chemicals or temperature, but with enhanced chloride removal in the upstream desalter. By far the most effective means of reducing chlorides in the desalter effluent is double desalting (i.e., two-stage desalting). This ought to remove 90% of the $MgCl_2$ and 95% of the $CaCl_2$, both of which are subject to hydrolysis:

$$MgCl_2 + H_2O \rightarrow MgOH + HCl$$
$$CaCl_2 + H_2O \rightarrow CaOH + HCl$$

Other methods to reduce chloride escape from the crude unit desalter which I have found effective are:

- Reduce the mix valve delta P to reduce the emulsion layer carry-over with crude.
- Avoid NaOH in the wash water.
- Adjust the wash water pH to achieve a 6½ to 7.0 of the wash water effluent brine. This may sometimes require addition of H_2SO_4 to the desalter.
- Maintain temperature about 240°F (for lighter crudes) and 290°F (for heavier crudes). Excessive temperature will reduce the voltage that can be obtained in the desalter grids.

- Add NaOH to the desalter effluent, up to a maximum of three or four ppm on crude charge rate.
- Keep the desalter pressure well above the calculated vapor pressure of the crude (assuming saturation with water).
- Provide sludge busting steam in the bottom of the desalter.
- Maintain the emulsion level in the desalter as low as possible, consistent with not increasing the crude oil content of the desalter effluent brine to the refinery effluent treating plant.

Salt contents of less than 3 ppm of crude are excellent. Above 15 ppm is poor desalting.

10

Pressure Control of Vacuum Towers

My wife, Liz, has just shown me an email announcing that the Hovensa Refinery in St. Croix is shutting down in February 2012. After reading this absolutely true chapter, I think you will see why. I wrote this chapter several months before the demise of this giant and relatively modern facility. It all proves what I have always thought. The key to refinery profitability is tech service. Even the part about the asparagus is true. The names have been changed to protect the guilty who lost $400 million per year from 2009–2011.

I recall standing in the control room of the Hovensa Refinery in St. Croix. Thomas, the young unit engineer, was in a rush to explain.

"Mr. Lieberman, I've only got a few minutes. I have to leave by 4:00. PRC-124 is the vacuum tower top pressure controller. As you can see, the control loop needs to be tuned."

The flash zone pressure was swinging between 20 mm and 26 mm of Hg. Not only was the vacuum erratic, but it was erratically bad.

"What's the rush?" I asked.

"Tonight's my girlfriend's birthday. I'm cooking her fresh asparagus, which takes quite a while. I'm quite a cook," Thomas beamed proudly.

"Okay, but where is the flash zone pressure indicator for the vacuum tower? Where is PRC-124 in the field?" I asked.

Thomas glanced at the wall clock, "Let me show you quickly. It's getting a bit late."

Outside it was really hot and really humid. Inside my Nomex coveralls, I started to slowly melt.

We climbed flight after flight of steps. Six landings up, we passed the vacuum tower flash zone. But Thomas continued up and up, until nine flights later, we stood on the very top of the vacuum tower. I could see the beautiful blue Caribbean Sea lined by the white broken coral beach. At this height, a cool sea breeze was blowing.

Thomas pointed to the spillback control valve (see Figure 10.1), sited on the top deck.

"There's PRC-124, the flash zone pressure indicator. Mr. Lieberman, I've really got to go now."

"What?! Where?! That's a damn control valve, not a pressure sensing point. We're at the top of the tower. And where's the pressure sensing control point for this valve?"

Thomas glanced at his watch and said, "I really need to leave soon."

"Anyway," I said, "Look at the spillback valve. It's going from full-open to full-closed every two minutes. That's what's causing the tower pressure to swing from twenty to twenty-five millimeters of mercury."

"Yes, I know. That's what is causing the tower pressure instability. The control loop needs to be tuned," Thomas said, screaming to be heard above the nearby roaring first stage ejector, which was surging.

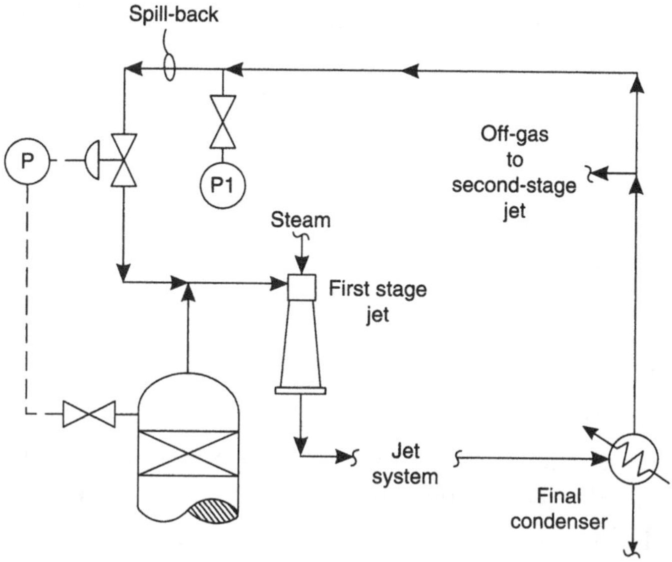

Figure 10.1 Spill-back value opening caused jet to surge. Thus, the pressure control value promoted rather than reduced tower pressure instability.

I had placed my digital vacuum gauge on the bleeder valve (P-1) upstream of the spillback valve. Here's what I observed:

- The spillback valve would open to 100%.
- The tower top pressure, as indicated by my digital gauge, would jump by ten mm of Hg.
- The first stage jet, located a few yards away, began surging, a sure indication that it was losing its sonic boost (see Chapter 1).
- The spillback valve would shut.
- The jet would stop surging, indicating that the sonic boost was restored.
- The tower top pressure would drop by ten mm of Hg.
- The spillback valve would open to 100%, and the cycle would be repeated every two minutes.

"Thomas, this is just plain stupid," I said, as I blocked in the spillback valve.

"Mr. Lieberman, you can't do that. We'll have no pressure control." Thomas had forgotten all about the asparagus. "We'll get into trouble."

"It's an experiment. I'll open it back up later. Stop worrying and find me the flash zone pressure sensing point."

"Wait, Mr. Lieberman. I just remembered where it's at. I'll show you."

We climbed down seven flights of stairs. Thomas pointed to a pressure transmitter labeled PRC-124.

"Thomas, that's not the flash zone either. It's above the wash oil grid (see Figure 10.2). The flash zone is two landings down, below the wash oil grid. But let's check the pressure here anyway with my digital vacuum gauge."

My gauge indication, above the wash oil grid, was a steady 18 mm of Hg.

"See Thomas, how steady the pressure is? The cause of your pressure instability was the first stage jet surging. Every time the spillback valve opened, it would overload the jet. The jet would surge and lose its sonic boost and half of its compression ratio. The loss of compression would cause the tower top pressure to jump, which would close PRC-124, unload the ejector, and stop the surging, thus restoring the jet's sonic boost (Figure 10.1). You understand?"

"But, Mr. Lieberman, where's the flash zone pressure transmitter?" Thomas asked.

"There isn't one. But we'll check the flash zone pressure locally with my digital gauge."

"I can't, Mr. Lieberman. I'm leaving now. I'm cooking lobster, too, for my girlfriend's birthday."

And would you believe it? When I finally did check the flash zone pressure (from the top of the over-flash pan level transmitter ¾ inch vent) (Figure 10.2), I observed a pressure of seventy (70) mm of Hg. Unbelievable! The wash oil grid

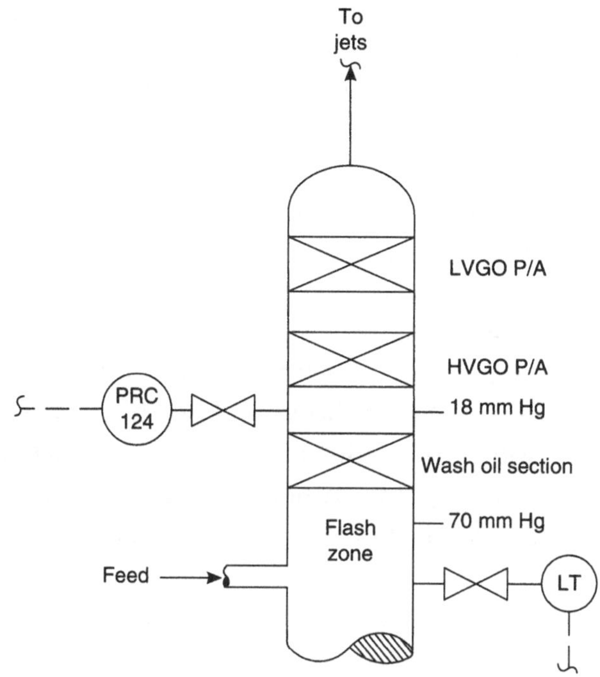

Figure 10.2 PRC-124 was not the correct flash zone pressure indicator location point. Note the huge wash oil grid delta p of 52 mm Hg.

had a delta P of 50 mm of Hg. A normal delta P should be one to four mm of Hg.

And when I reported this fiasco on Friday to the plant management staff, Mr. Grimson, the technical services manager, turned to Thomas and asked, "How did the dinner turn out? I love fresh asparagus when they're cooked just right."

Okay. I made up the part about Mr. Grimson. But the rest of the story is absolutely true. Even the asparagus part. Can you just imagine how badly the wash oil grid must have been coked to develop 50 mm of Hg pressure drop?

I have shown (Figure 3.2 pg. 30, in my book, *Process Design for Reliable Operations*) a photo of a wash oil grid for a vacuum tower. Actually, this is for a very small (about six feet I.D.) tower. The packed height (about three to four feet) is normal. Refinery vacuum towers that I have worked on vary from 12 to 46 feet in diameter. However, regardless of the tower I.D., increasing the grid height much above four feet serves no process purpose and makes the grid more prone to coking. I can't imagine how the Hovensa Refinery in this story developed a 50 mm Hg delta P across the wash oil grid and was unaware of the problem. If I hadn't seen it myself, I would never have believed the story.

Another pressure control method is used by the Texaco (now Motiva) Refinery on their giant lube oil vacuum tower in Port Arthur, Texas. The jet shown in Figure 10.3 was designed to develop a sonic boost. But the jet had not worked with its sonic boost for a very long time. As long as an ejector is only developing its velocity boost, it is said to be in its "throttling

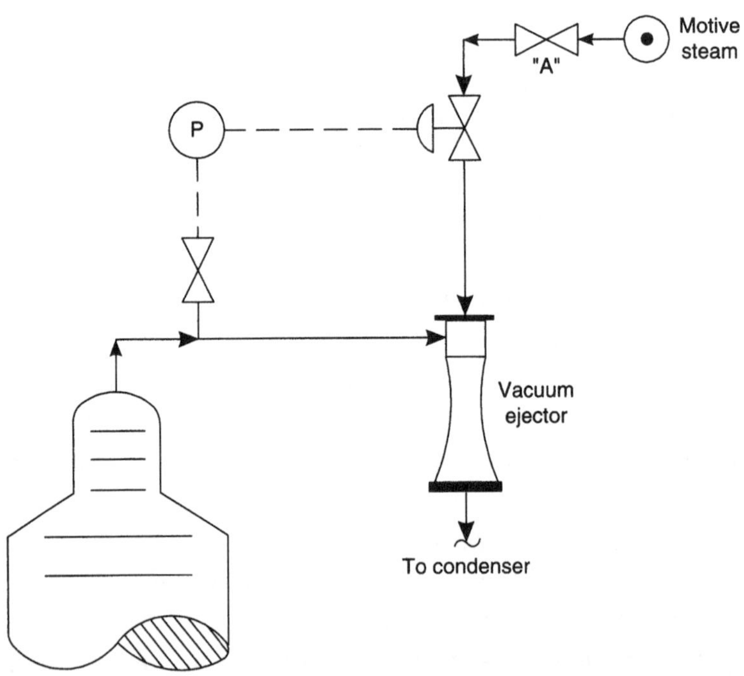

Figure 10.3 Throtting on motive steam supply sometimes works well in vacuum tower pressure control, but sometimes causes a sudden break in the vacuum if the sonic boost is lost.

mode of operation". This means that a small closure of the motive steam inlet control valve will result in a small reduction (i.e., increase in pressure) in the tower vacuum. Increasing the motive steam pressure did improve vacuum, but only to a point, after which the vacuum became worse.

10.1 Positive Feedback Loop

Even when operating in the throttling mode (without the sonic boost), too much steam pressure can sometimes reduce the jet's compression ratio. This may happen because:

- The downstream condenser is overloaded.
- The steam nozzle in Figure (1.1) is worn out because of wet steam or lack of maintenance.
- The motive steam pressure is above its design pressure.

11

Condenser Cooling Water Flow

11.1 Cooling Water Flow Configuration

I'll elaborate on the example that I introduced in Chapter 4. It illustrates the importance of avoiding design of process equipment without considering the individual circumstances.

The largest vacuum tower pre-condensers I ever saw were the eight giant exchangers at the former Esso Largo Refinery in Aruba. I have sketched the nozzle and tube support baffle configuration in Figure 11.1. Can you see Esso's mistake? It required nine years and fifty-odd trips to Aruba for me to notice the design malfunction.

First, observe that there is no air baffle in the condenser (see Figure 11.2). Also, note that the vapor outlet is on top of the exchanger, not on the side. The cooling water outlet, and therefore, the final rows of tubes are also on top. This means that the warmest cooling water is contacting the condenser vapor effluent.

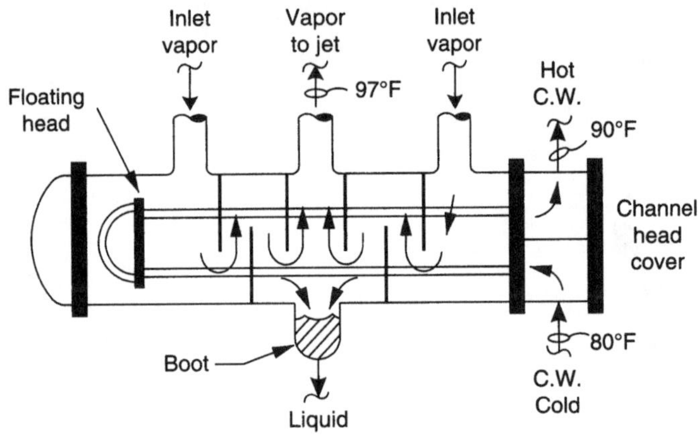

Figure 11.1 Reversing cooling water flow to bring the cool water into top of channel head dropped vacuum tower flash zone pressure.

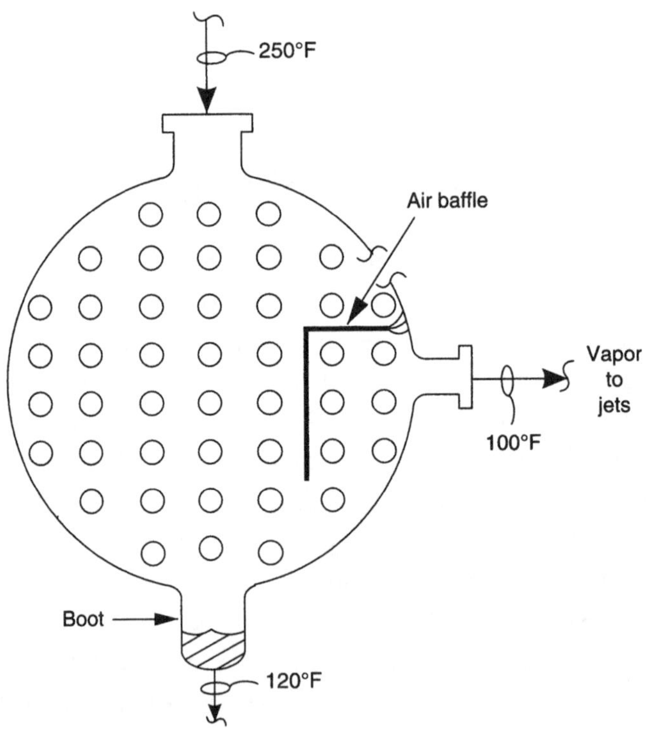

Figure 11.2 Air baffle minimizes moisture content of vapor outlet to jets. Vapor should be 20°F–30°F colder than boot.

It's true that for water coolers in general, it's best to have the cold water entering the bottom of the channel head, and the warm water leaving at the top of the channel head, as shown in Figure 11.1. But in this condenser, the effluent vapor, the temperature which we are trying to minimize, is also exiting from the top of the exchanger shell. So, I had the cooling water flow reversed (i.e., the opposite as shown in Figure 11.1) in the channel head of all eight shells by a relatively simple change in the cooling water supply and return piping headers. The result of this reversal in water flow direction was to reduce the pre-condenser outlet temperature by an average of about 2°F to 3°F. This resulted in a reduction of the vacuum tower flash zone pressure of about three to four mm of Hg, Meaning the flash zone pressure (see Table 4.1) dropped by an average of four to five percent. This further resulted in a significant increase in gas oil recovery from the vacuum tower bottom's residual tar (i.e., coker feed).

One could argue that a better solution to this design malfunction would be to retrofit the pre-condensers with air baffles and side draw-off nozzles for the vapor (Figure 11.2). After all, that's the standard and correct engineering design for pre-condensers. But that would have required new bundles and new nozzles. My client, Coastal, would never have approved such a major project. But, that's the nature of working with process equipment suffering from design errors. You have to do the best with what you have, and learn from other engineers' mistakes.

11.2 Air Evolving from Cooling Water Reduces Cooling Water Flow

I have learned a lot about how process equipment actually works by investigating comments such as, "It may not make any sense to you, but that is what happens here."

Process equipment, however, always conforms to the principles of science, but we have got to know which principle to apply.

Figure 11.3 shows an old vacuum tower in Aruba. The chief operator on this unit made the following statements:

- "The colder the vapor outlet temperature from the pre-condenser, the better the vacuum could pull, because of reduced vapor flow to the jet." Agreed!
- "Increasing cooling-water flow to the pre-condenser decreases the vapor outlet temperature." Agreed!
- "Closing the cooling water outlet valve A, (Figure 11.3) about three-fourths of the gate valve stem travel, increases cooling-water flow through the pre-condenser." Nonsense!

Really, dear reader, how can closing a valve in a pipeline increase flow in that pipeline? It cannot, and it will not, and it did not. Yet, on the other hand, it is a really bad idea to disregard field observations made by experienced plant operators. So, let's take a closer look at Figure 11.3.

First, I tried opening valve A. Just as the chief operator had said, the cooling-water outlet temperature increased, proving that water flow was reduced. Next, I checked the pressure at bleeder B; it was 12 inches of Hg (i.e., 460 mm of Hg absolute). The pressure was so low at this point because of:

- The 20 psi ΔP of the cooling water as it flowed through the tube bundle.
- The 35 ft of elevation (about 15 psi of head pressure) that the water had to gain to climb to valve B. Of course, this 15 psi of head loss was regained when the water flowed back down to the cooling-water return header. (This was a water return channel that operated at atmospheric pressure.)

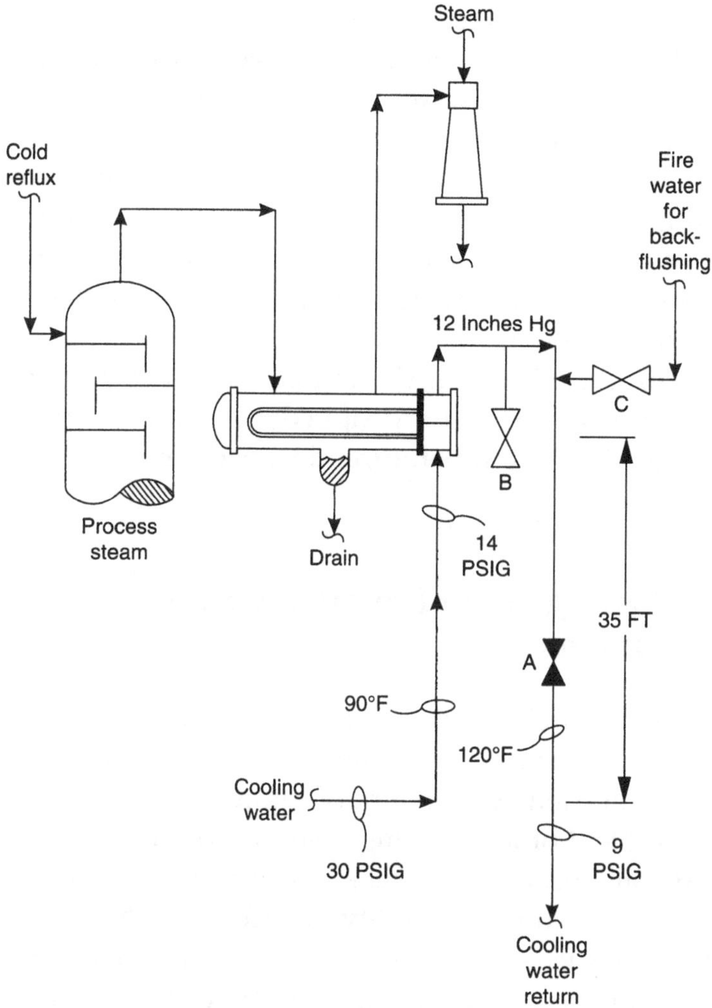

Figure 11.3 Air evoloving from cooling water reduces water flow.

The first idea I had was that the 120°F water would partially flash to steam at 12 in Hg and the evolved vapor would restrict water flow. Wrong! The vapor pressure of water at 120°F is 26 in Hg, not 12 in Hg.

But cooling water is not just pure water. It is water that has been saturated with air in the cooling tower. Sure enough, when I calculated the amount of dissolved air that would flash out of water (which had been saturated with air at 90°F and atmospheric pressure), I found that a very large volume of air could flash out of the water. It was just as the chief operator had said. Opening valve A too much, evolved large volumes of air in the second pass of the U-tube pre-condensers, shown in Figure 11.3. Certainly, the total volumetric flow through the outlet of the condenser does increase, as valve A is opened. But the incremental flow is all air, and all that air does is to choke off the cooling-water flow.

As I throttled back on the outlet water flow, the cooling water outlet temperature dropped by a few degrees. The condenser outlet temperature declined, and the vacuum improved a bit.

11.3 Cooling Water Pressure to Surface Condensers

I learned this trick of throttling on the cooling water supply flow in the Aruba Refinery from Leo Chago, shortly before he passed on. A friend lost, but not forgotten.

The problem of low cooling water pressure to the elevated surface condensers may cause two other types of malfunctions: air leaks and the inability to back-flush the tubes.

If a valve is leaking on the cooling water return line, at the elevation of the condenser, air may be continuously drawn into the flowing water. As in the previous example, the air flow reduces the flow of water by displacing some water flow from the return water piping. If you think a fitting or flange is leaking, spray a water hose over the suspected leak. Water, not air, will be sucked into the leak. If the cooling water outlet temperature from the condenser drops, you have located the leak.

The ability to back-flush cooling water tubes in all services is critical. Wood chips, sand, and shells (but not calcium hardness deposits) can be removed by flowing water from the cooling water return header backwards through the tube bundle, and draining the water to the sewer. Of course, this is quite impossible if the return header pressure at the elevation of the surface condenser is below atmospheric pressure.

We had this problem in Aruba on eight giant vacuum tower overhead pre-condenser shells. To provide high pressure water for routine back-flushing, we tied a ten inch line from the refinery firewater circuit into a new back-flush header, above the eight condensers. The firewater pressure was sufficient to back-flush the condensers. That was in 1998. Now in 2012, I'm older and wiser. Now, I would not use plant firewater for such a routine process service. Perhaps in an emergency, I would not have sufficient pressure in the firewater system to control a fire.

11.4 Tube Leaks

I never had a tube leak in a surface condenser. But, when I checked for one, the procedure was pretty simple. I would get a sample of steam condensate from the condensate recovery pump. As we used chromates in our cooling water treatment program, I would just have our lab check for chromates in the condensate. The concentration of the chromates would then correlate with the magnitude of the exchanger leak. I do not believe that a relatively small water leak could seriously impair the efficiency of the surface condenser. Your cooling water chemical vendor will have a portable test kit that can be used for whatever treatment chemical used in your cooling water system.

If the water leak is large, and results in liquid backing-up into the condenser, then I would imagine that the resulting loss in vacuum might then be quite significant.

12

Condensate Back-Up in Condensers

One day, an energy consultant expert visited the former Gulf Refinery in Port Arthur, Texas. He advised Gulf's management that in order to save steam, the surface condenser outlet temperature connected to condensing steam turbines should not exceed 100°F. This limitation was then conveyed to the plant operators as a, "Standard Operating Instruction."

But there was a problem. The cooling water supply temperature in Port Arthur in July is around 93°F. This makes it rather difficult to reduce the surface condenser vapor outlet temperature to 100°F. However, the surface condenser outlet temperature, for every steam turbine that I have seen, is not measured at the vapor outlet, but in the hot well or boot, as per Figure 12.1. This measures the steam condensate drain temperature, rather than the vapor temperature to the downstream jets.

So, inadvertently, the Port Arthur operators had been instructed to minimize the boot temperature, rather than the

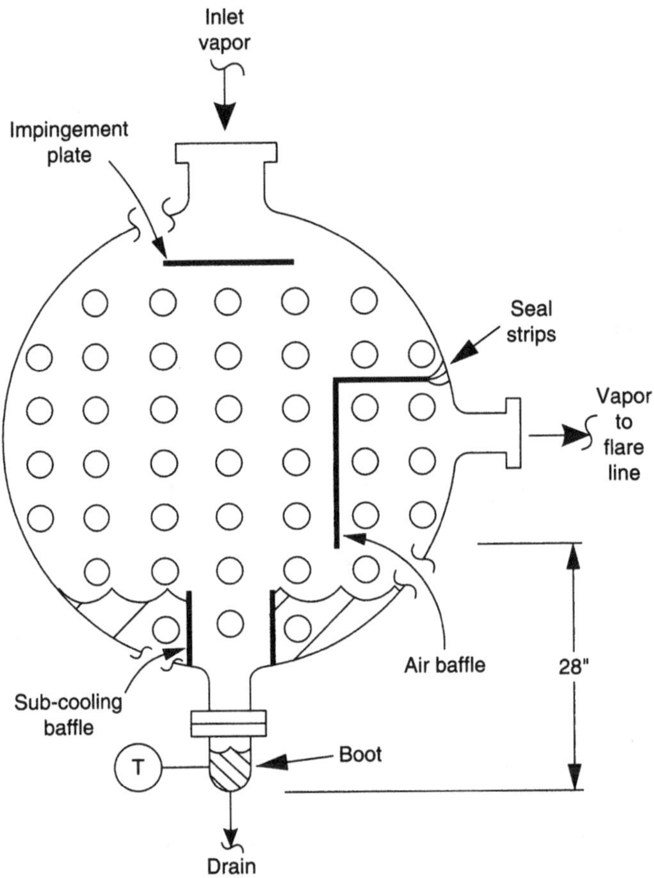

Figure 12.1 Internal components of a vacuum suface condenser. Seal strips retard vapor leakage around the air baffle.

flowing vapor temperature, to the jets. And, that's exactly what they did. They backed the steam condensate level in the boot up into the surface condenser itself. This promoted a greater degree of sub-cooling of the condensate. It also promoted:

- A reduction in surface area exposed to the condensing vapors.
- Which raised the vapor outlet temperature and the vapor flow to the downstream jets.

- Which caused the pressure in the surface condenser to rise.
- Which reduced the amount of work that could be extracted from each pound of steam.
- Which then required more steam flow to drive the turbine.

Thus, steam was wasted rather than saved by giving the operators an arbitrary guideline for the surface condenser outlet temperature. And who's fault was this? Well, it just illustrates the importance of having knowledgeable technicians or engineers working with the operators in the field to accomplish management objectives.

Half the problems that we have with poor vacuum in the surface condensers are not related to the condensers themselves, but to downstream malfunctions in the steam jets or ejectors.

And, it's the surface condenser vapor outlet temperature and certainly not the boot or hot well temperature that controls the moisture content of the vapor that overloads the downstream ejector system.

12.1 Undersized Condenser Drain Nozzle

The vacuum tower at the Calumet Refinery in Shreveport, Louisiana, would develop a reasonable vacuum at low feed rates. As feed rates increased, the tower top vacuum became slightly worse, until at design rates, it suddenly broke and went off the chart. Only one of the two first stage jets was in service. Unit operators stated use of both jets did not help. My visit happened to coincide with one of the coldest periods in the history of northern Louisiana. Hence, the condensers' cooling water supply temperature was very low.

As this story occurred in late 2010, I can easily recall running my fingers along the exterior of the pre-condenser shell and noting a temperature change about 30% above the bottom of the shell. The pre-condenser was suffering from condensate back-up. The lower portion of the condenser was cold and the upper 80% was warm. The condenser boot and entire liquid outlet line was equally cold. Two things puzzled me:

First, why all the condensate back-up? Shouldn't the liquid level be down in the boot (see Figure 12.1), rather than backed-up into the condenser?

Second, most of the tubes in the condenser were still not covered in liquid. Considering the very low cooling water temperature, condenser capacity ought to be adequate anyway, even with a modest degree of condensate back-up.

However, I then noted that there was a piping restriction on the condensate drain line. While both the boot size and the drain line were sized properly, the outlet nozzle from the boot was smaller than the drain line. I measured the height of the boot, plus the height of the condensate back-up in the condenser at 24 inches.

Next, based on the water and hydrocarbon production from the seal drum (both flows were metered), I calculated the volume and velocity of flow through the small boot drain nozzle. This velocity was 8 ft./second. Using the nozzle exit loss equation:

- $\Delta H = 0.34 \bullet V^2$
- $\Delta H = 0.34(8)^2 = 23''$ of liquid

As my calculated level of condensate back-up corresponded closely with my measured height of back-up, I now had accounted for the observed liquid level in the pre-condenser.

Now the plant manager of the Calumet Refinery, Tom Germany, had explained to me that the minor loss in vacuum observed, as feed rates increased was not his real concern. Rather, it was the sudden breaking of the vacuum, at a feed rate less than design.

"Norm," Tom explained, "This is bottlenecking our entire operation."

Why then the sudden loss in vacuum at a particular feed rate? The explanation is, like all our work, a bit complex:

- Increasing feed by 10%, increased flow through the pre-condenser drain nozzle by 10%.
- The 10% higher nozzle velocity, increased condensate back-up by about 20% (ΔH or ΔP varies with velocity squared).
- Meaning the liquid level in the condenser was pushed up by another 20% or about five inches (20% × 24″ = 5″).
- However, as can be seen from Figure 12.1, this pushed the liquid level up to submerge the bottom edge of the air baffle, which was 28 inches above the boot outlet drain nozzle.

Covering the air baffle inlet with liquid would result in massive entrainment of liquid to the first stage ejector and the vacuum breaking. At any rate, we shut the vacuum tower down, enlarged the boot outlet nozzle and the unit charge design rate was achieved without further problems.

12.2 Seal Drum Level Indication

Whenever the condensate level in any condenser backs-up to submerge the air or vapor baffle, the vacuum tower pressure is going to break. For example, after we fixed the problem with the restrictive boot nozzle at the Calumet vacuum tower, we began to experience the same symptoms. Now what?

I noted two new indications of a problem: One was that the seal drum level (see Figure 12.2) always read 70%. Not 71% or 69%, but 70% – for weeks. Second, the level control valve on the condensate pump removing liquid from the seal drum was 100% open with its by-pass wide open.

I blew out the level taps on the seal drum level glass, and found that the drum level was really 100% plus. The flooded

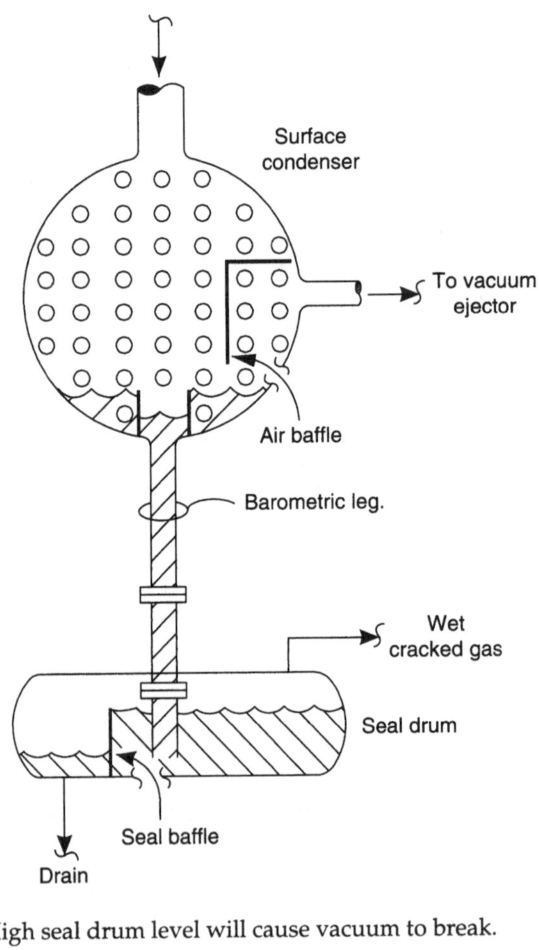

Figure 12.2 High seal drum level will cause vacuum to break.

seal drum was backing liquid up in the pre-condenser again. What had happened?

The primary condensate pump had been removed from service for a mechanical repair. The spare pump was undersized and could not handle the flow from the seal drum. The operators, trying to maintain a 50% level set-point in the drum, opened the by-pass around the vessel's LRC. The control panel level lined-out at 70%, even though the seal drum was 100% full. The level indication calibration was faulty, hence the discrepancy.

I had the larger pump retrieved from the repair shop, the seal drum level indicator recalibrated, and all was well again with Tom Germany's vacuum tower pressure in Shreveport, Louisiana.

12.3 Leaking Gauge Glass on Surface Condenser Boot

The Good Hope Refinery had a string of seven turbine driven air blowers supplying combustion air to their catalytic cracker regenerator. All seven exhausted their motive steam to surface condensers. I had become the technical service manager of this, the world's largest privately owned independent refinery, in December, 1980. Although I had worked as a refinery process engineer since 1965 for Amoco Oil, I really did understand how a surface condenser worked.

Figure 12.3 shows a typical surface condenser configuration. The main points I would like to make about this sketch are:

- The level in the boot (normally called the hot well) is controlled beneath the bottom row of tubes.
- The temperature of the boot and vapor outlet will be the same, within a few degrees, provided the tubes are not covered by condensate.

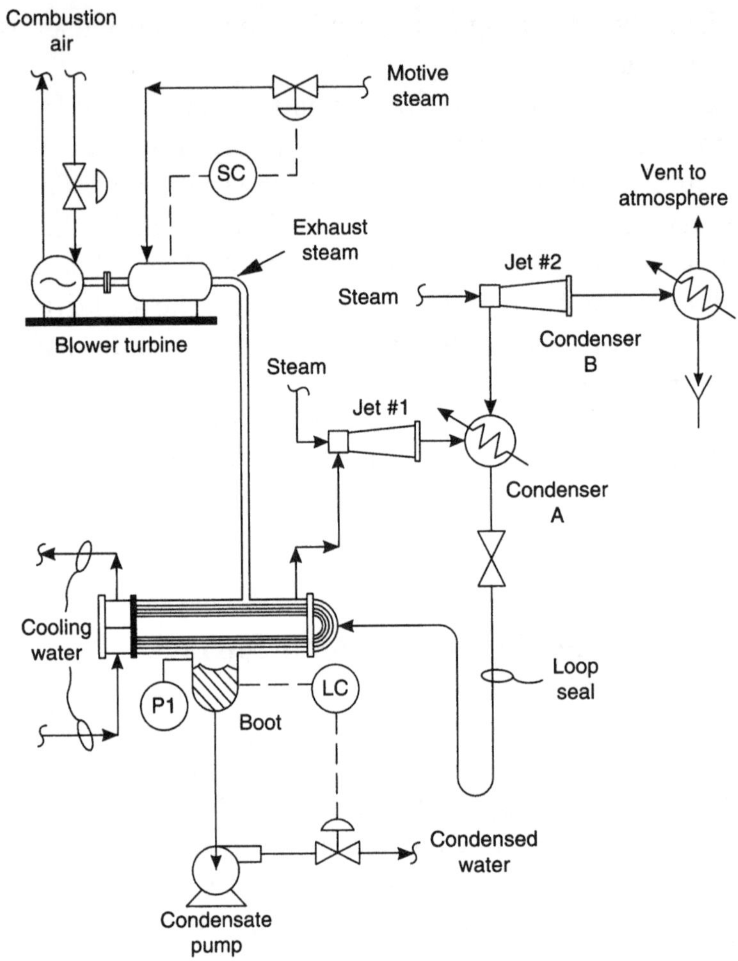

Figure 12.3 Condensing steam turbine with standard two-stage ejector system driving an air compressor.

- The condenser outlet temperature is almost always measured only in the boot.

The problem with K-701, one of the medium sized air blowers, was that it was running too slow. The operators had reported this to Frank Elvin, the plant manager. The problem was defined to me by Mr. Elvin as follows:

- The operating pressure of the surface condenser was 175 mm of Hg (23 inches of Hg vacuum). It used to be 100 mm of Hg (26 inches of Hg).
- Due to the higher condensing pressure, less work could be extracted from each pound of motive steam.
- Hence, the turbine and air blower were running slow.
- The reduced turbine speed reduced the polytrophic head developed by the air blower.
- The reduced head developed by the blower, reduced the blower discharge pressure below the cracker's regenerator pressure, and thus precluded the use of combustion air from the blower.

"So your job, Norman, is to improve the surface condenser vacuum from 23 inches of Hg to 26 inches of Hg," concluded Frank Elvin.

I'm really not much of a manager. I had a dozen engineers reporting to me at the Good Hope Refinery, but I have never mastered the concept of delegating problem solving. So in 1980, and even today at age 70, my only response to defining a malfunction is to grab my wrench, fittings, and gauges, and tackle the problem in the field.

I could see that the water level in the boot level glass was clear and steady at 50%. As the level-trol (used to measure the boot level) was connected to the same set of taps as the glass, on the boot, the level that was transmitted to the level control valve (LRC), must also be about 50%.

But then, I noticed something really odd. The boot temperature was cool to the touch. The local dial thermometer was reading 88°F. But the condenser vapor outlet going to the two-stage vacuum system was quite warm. I checked this temperature with my infrared digital temperature gun and it read 132°F. Why the 44°F difference?

At 132°F, the vapor pressure of water is almost 175 mm of Hg (see Table 4.1) (i.e., 23 inches of Hg vacuum). And, it is the surface condenser vapor outlet temperature that controls the condensing pressure, not the hot well or boot temperature.

The water in the boot was obviously sub-cooled. But, why? Certainly this was confusing. For lack of anything better to do, I decided to place my vacuum pressure gauge on the boot's gauge glass (P-1 shown in Figure 12.3), just to get a local reading of the surface condenser boot's pressure.

Installation of my pressure gauge was easy. The screwed ¾" connection on top of the gauge glass didn't even have a bull plug. It just screwed into a ½" to ¾" bushing and then my vacuum pressure gauge. And then my world changed. Even before I opened the valve to get the pressure reading, the pressure gauge needle was drawn down to read 23 inches of Hg vacuum. Obviously, the top level tap connection at P-1 was leaking. However, screwing in my gauge stopped this air leak.

The air leak had increased the pressure at the top of the gauge glass and level-trol. This had pushed down the level in both the gauge glass and level-trol. And, as a consequence:

- The level transmitted to the LRC valve was reduced.
- The level control valve started to close.
- The boot level rose.
- The condensate level was pushed up into the condenser itself, and the heat transfer tubes were submerged in water.
- The rate of steam condensation slowed, and the condenser vapor outlet temperature increased.
- The condenser pressure rose and the steam turbine and blower slowed.

When I stopped this leak with my pressure gauge:

- The indicated boot level rose.
- The boot LRC opened.

- The condensate level dropped out of the shell exposing more tubes to the turbine exhaust steam.
- The condenser pressure dropped and the blower speed increased.

The air blower discharge could now be lined up to the cracker catalyst regenerator. Frank Elvin was really pleased. He said that henceforth he would direct operations in the east side of the refinery and I would direct operations in the west side of the plant. Just one week after I joined the Good Hope Refinery, I had advanced more in my career than I had in the previous 16 years at Amoco Oil. For the first time in my adult life, I had a sense of belonging and fulfillment.

12.4 Condensate Pump Cavitation Due to Air Leaks

Both of the following two malfunctions I experienced at the visbreaker vacuum tower at the Good Hope Refinery. However, I solved both problems in just a few minutes because I had previously also experienced both problems at home on my swimming pool water circulation pump.

My pool pump sits on a raised concrete slab about 2½ feet above the water level in my pool. The pump suction pressure is therefore about two inches of Hg vacuum (i.e., minus one psi below atmospheric pressure). The plastic (PVC) suction piping connections are old and subject to leakage. Air is drawn into the pump suction through these leaks. I can see the air as it bubbles through the transparent pump leaf basket. A small air leak can cut the pool water circulation rate by 20%–30%. To identify the exact connection that is leaking, I'll run water from a hose over each connection. When I drown the leaking connection, the pump discharge pressure will jump from 10 psig to 16–20 psig.

Of course, I'm too busy to actually fix the bad connection. I'll just wrap it with grey duct tape which lasts a year or two.

The other problem I can't fix with duct tape is a leak on the pump's mechanical seal. The pressure inside the pump case is not the pump discharge pressure, but the pump suction pressure. As a matter of fact, the entire pump case is slightly lower than the pump suction pressure. Hence, a seal leak draws air into the pump case and greatly diminishes the pump capacity and discharge pressure. Again, I'll run a garden hose over the pump's seal. If the pump discharge pressure then jumps by five or ten psi, I have a seal leak. To change out the rubber seal is a multi-hour job.

At the Good Hope Refinery, the loss of pump discharge pressure due to these sorts of air leaks resulted in a high boot level and consequent loss of the surface condenser vacuum.

The air leaks were both identified with a water hose. One such air leak was on the packing gland of a gate valve on the pump suction. I tightened both packing follower nuts on top of the valve to stop the air leak identified with the hose.

The second leak was on the pump's mechanical seal. I left the hose running across this seal, as the spare pump was lost in the maintenance pump non-repair department. As soon as the water ran across the seal, the pump discharge pressure jumped up, the boot level dropped back into the gauge glass, and the visbreaker vacuum tower flash zone pressure was drawn down in a most satisfactory fashion.

12.5 Condensate Back-Up in Surface Condenser Boot

"Look, Norm. We've got to get K-805 running. I've spent four million dollars to have it installed. And now it won't run fast enough. We need that air real bad," said Jack Stanley, the owner of Good Hope Refinery.

K-805 (Figure 12.3 page 158) was an old used air blower that Jack had purchased for $20,000 on the used equipment market. It was needed to supplement the combustion air supply to our FCU (fluid catalytic cracking unit) catalyst regenerator.

"Jack, it runs," I explained, "But not fast enough to put up the 45 psig of air pressure, which we need to get into the spent catalyst regenerator."

Mr. Stanley studied his elongated fingers for a minute. "What's the problem Norm? Not enough steam pressure?"

"No, Jack. Too high a back pressure from the surface condenser. It's running at 300 mm of Hg (i.e., 18 inches of Hg vacuum). It's supposed to be at 100 mm of Hg (26 inches of Hg vacuum)."

"Maybe the cooling water's not cold enough," Jack said.

"Nope. The cooling water's fine," I answered.

"Could be the condenser tubes are plugged. Maybe we're not getting enough water flow through the tubes. Maybe the channel head inlet is restricted?"

"No, Jack. I've dropped off all the channel head covers (i.e., inspection doors used to clean and inspect the channel head tube sheet on the water side of the surface condenser). It's all clear."

"Could be the boot level's too high. Steam condensate is backing up and covering the condenser tubes and retarding the rate of steam condensation. Norm, that's happened before on the other surface condenser," Jack observed.

"Yeah, Jack. I know. But I can see the damn level in the boot's gauge glass. It seems to be okay. The funny thing is the boot is cold. Considering the high pressure that it's running at, it ought to be hot."

"I'm confused, Norm. Why should the boot be hot, if the vacuum's bad?" asked Mr. Stanley.

I ought to explain something about Jack Stanley. He had borrowed $1.3 billion in 1980 from some banks to build the Good Hope Refinery, and he was unable to repay the loan.

Jack did not have any technical training, but understood most branches of engineering reasonably well. He was dictatorial, arbitrary, and mean to his employees and family. I loved working for him and miss him to this day.

"Jack, the turbine exhaust steam driving the air blower condenses at a certain pressure. The lower, the better. If steam condenses at a cold temperature, it ought to condense at a lower pressure. The boot temperature should represent that condensation temperature. At the current cool boot temperature, the steam should be condensing at 100 mm of Hg, not the actual 300 mm of Hg. If I could get the condenser pressure down to 100 mm of Hg, I could bring the blower up to speed and generate the needed air blower discharge pressure."

"Look, Norm. I've got to have dinner with my mother tonight. See if you can't get this problem fixed by the morning," Jack said, as he picked up the keys to his blue X-12 Jaguar.

Now what?! The sun had set across the Mississippi River. The engineering and administration staff had all left. It was a dark, hot, humid night. My girlfriend, Linda, was waiting for me to take her to dinner in the French Quarter. I wandered over to look at K-805 once again.

The maintenance crew assigned to assist me, stood around. They were cheerful and chatty. Why not? They were being paid time and a half for doing nothing. It was then that I noticed that the condenser boot (see Figure 12.1) was not welded, but bolted onto the condenser shell.

"Mr. Craig," I asked the old maintenance foreman, "Why is it that the boot bolted up to the shell?"

"Lieberman, we don' have that kinda condenser in the Navy. Need to drop off that there boot once in a while to clean the crap offa them bottom few rows of tubes. Them tubes get crud-ed up with silicates and salts and stuff. Water can't get into the boot. Backs up and ruins the vacuum. Then, the turbine's gonna run slow."

"Craig, how long will it take you all to drop the boot, hydroblast the outside of the tubes and bolt the boot back up?"

"Eight hours."

"And if I sign your time cards that you worked eight hours, how long will it take?"

"Two hours. Maybe, three."

"Okay. I'll be back at nine. Let's get rolling," I said as I headed for the plant gate.

I dropped Linda off at her decaying house by the levee after dinner, and then drove back to the refinery. The boilermakers were just tightening the last bolts on the boot.

"Ready to go, Mr. Lieberman," Craig said, as he handed me the five falsified time cards to sign.

The night shift operators on the FCU plant opened the 120 psig steam valve to the turbine, lined up the condensate pump, and the giant K-805 air blower roared to life. And roar it did, as the surface condenser vacuum fell happily to 110 mm of Hg. I had the operators close the air vent and open the air valve to the FCU regenerator.

The next morning I explained the successful commissioning of K-805 to Mr. Stanley.

"Amazing Norm," he said, "That Craig's maintenance crew did all that work in only eight hours."

"Yes Mr. Stanley," I agreed, "They were really amazing."

12.6 Experiment with Condensate Back-Up

Often my clients think I'm conducting field observations on their process facilities. But in reality, I'm playing with their equipment. For example, I was working in a refinery in Arkansas City, Texas, on a vacuum tower problem. I found that I could cause the vacuum to break by increasing the level in the seal drum (see Figure 12.2):

- Raising the level in the seal drum backed liquid up into the condenser.

- This caused the liquid drain temperature to go down, but the condenser vapor outlet temperature to increase.
- The high vapor outlet temperature increased the vapor pressure of the water in the condenser.
- This increased the discharge pressure of the upstream ejector above its "Critical Discharge Pressure." This critical pressure is shown on the jet system specification sheet issued by the manufacturer (Figure 3.1).
- The jet was forced out of its "Critical Mode of Operation." That means that the jet lost its sonic boost.
- The jet began to surge and the vacuum broke in the upstream vacuum tower.

Lowering the level in the seal drum allowed me to restore the sonic boost within a few minutes, and thus recover the vacuum. And here's the best part. My wife, Liz, delivered a sandwich to me on the 6th level platform, so I wouldn't have to interrupt my important field test for lunch.

12.7 Condensate Back-Up

Once, I was working with a jet with an observed compression ratio of about (180 mm Hg)/(150 mm Hg) = 1.20. This extremely low compression ratio did not indicate any sort of jet malfunction. The high jet suction pressure of 150 mm Hg, is caused by the 140°F pre-condenser outlet temperature. The vapor pressure of water at 140°F is 150 mm Hg. There was a large amount of process steam flowing into the pre-condenser. The lowest possible pressure that the pre-condenser can operate at and still condense the process steam is 150 mm Hg. As the jet sucks harder, it just pulls a few more pounds of water out of the pre-condenser, without altering the pre-condenser's pressure.

The problem with the pre-condenser was condensate back-up. Something, perhaps a partially plugged drain line, was restricting condensate flow. As the condensate backs up, it reduces the surface area of the condenser exposed to the condensing process steam. This makes it more difficult for the process steam to condense. The condensate back-up also subcools the condensate. The net result was that the pre-condenser vapor outlet temperature went up and the pre-condenser liquid outlet temperature goes down.

In Arkansas City, I was able to force a jet to surge and lose its sonic boost simply by raising the condensate level in its downstream condenser by just 6 inches. Lowering the level drew down the jet's discharge pressure by a few millimeters of mercury and restored the ejector's critical flow velocity and the ejector's sonic boost.

13

Seal Leg Drainage

I suppose the reader will be surprised that a minor subject such as seal leg drainage from a surface condenser would require a complete and rather long chapter by itself. However, this is a big problem that is rarely recognized. Even this extensive chapter barely scratches the surface of this really complex and thoroughly nasty subject. There is a very high probability, though, that the ultimate cause of your vacuum system problems is related to a seal leg drainage malfunction.

13.1 Sludge Accumulation in Seal Drum

In all the years I've worked in the identification of vacuum system malfunctions, I never encountered the problem shown in Figure 13.1. Never, until this week at the Sinclair Refinery, in Wyoming. Sludge in the bottom of the vacuum system's seal drum, restricted drainage from the vacuum inter-stage condensers. I obtained a sample of the black gritty material. It was plain iron sulfide.

Corrosion in the condensers, due to wet H_2S, plus a few ppm of HCl, is common. Such corrosion deposits will be flushed into the seal drum. Why I wondered, had I never encountered this malfunction before. Restricting condensate drainage will surely cause liquid back-up that floods the condenser. This raises the condenser pressure and causes the jets to bog down and the vacuum, as the operators say, "Will break!"

The reason for this unusual malfunction at the Sinclair plant, was that the bottom of the seal legs were only six inches above the bottom of the six food I.D. seal drum, as shown in Figure 13.1. As the overflow baffle was four feet high, the six inch clearance of the seal legs made no sense. The depth of the liquid seal only needs to be a few inches, not 42 inches (i.e., four feet minus six inches).

I suppose that the problem was aggravated by the failure of my client to wash out accumulated sludge from the seal drum during last year's turnaround, even though the plant operators

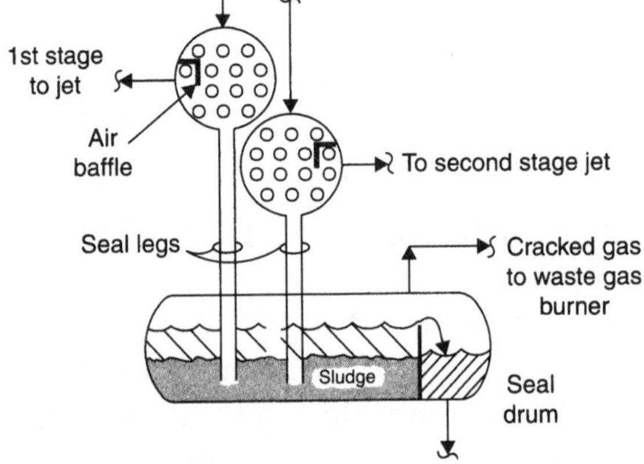

Figure 13.1 Sludge accumulation in a seal drum floods the condensers and causes the vacuum to break suddenly.

insisted that this was necessary. So now, when the vacuum breaks, the same operators must rush out to madly blow out the plugged seal legs with 200 psig steam.

At the next opportunity, I suggested that the bottom foot of the seal legs be cut off.

13.2 Seal Leg Leak Inside Seal Drum

Biological corrosion is a big problem when carbon steel piping comes into contact with moisture and H_2S at the warm temperatures we experience in vacuum system seal drums. The bacteria metabolize the carbon steel into iron sulfides. If O_2 is present, the problem is even worse. My experience has indicated that the internal seal leg piping exposed to the vapor phase in the seal drum is especially vulnerable to biological corrosion and holing-through. Hence, this piping is best constructed from high alloy steel like 316 (L) s.s.

At the Amoco Oil Refinery in Sugar Creek, Missouri, in 1978, I was involved in a serious incident which I accidentally caused. I had been sent to the plant to improve the performance of their vacuum tower. Mainly to reduce the tower top pressure. At this time, I was at the top of my career with Amoco and quite sure of myself.

"I see Joe," I explained to the old, tired looking chief operator, "that you're running with the seal drum level very high. Kindly reduce the hydrocarbon liquid level down below the top edge of the overflow baffle." (Figure 13.2)

Joe looked at me rather doubtful. "I don't know bout that Mr. Lieberman. Might not be a good idea."

"Look, Joe," I explained, "Lowering the level in the seal drum will help pull the level down out of the primary condenser that's suffering from condensate back-up. That vapor outlet is 130°F and the liquid outlet is 110°F. The liquids is subcooled and the vapor is too hot. The 130°F vapor is carrying a

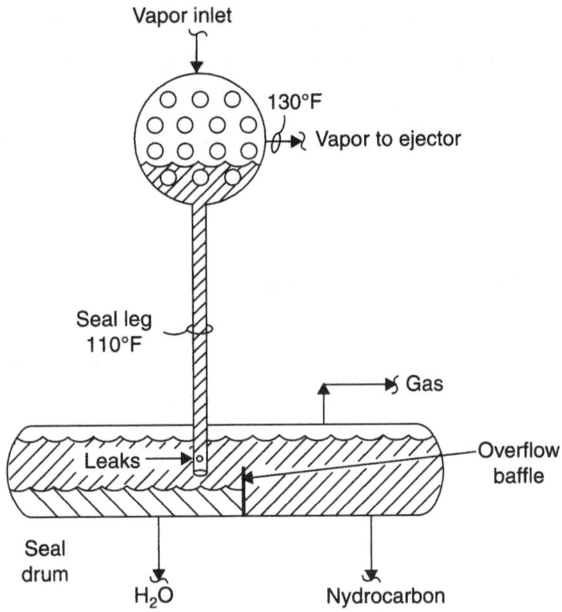

Figure 13.2 Hole in barometric drain leg inside seal drum forces operators to run seal drum flooded.

lot of moisture into the first ejector, which is being overloaded and losing vacuum."

Joe fidgeted in his ancient swivel chair. The brown leather was cracked with age. "I kinda don't want to lower that there seal drum liquid level, Mr. Lieberman. It's gonna break the vacuum."

"Why is that, Joe?"

"I surely don't know, sir."

"Well Joe. If you do not know, please follow my instructions. I'm here from the Chicago Engineering Division to improve refinery profits. Let's lower that seal drum level."

"Mr. Lieberman, I really would rather not do that. Somethin' bad might happen."

"Look Joe, I've got a Master's Degree in Chemical Engineering. I think you can rely on my judgment to make sound decisions. It's getting late. Please lower that level."

Old Joe surveyed the empty control room for help and support. Finding none, he sadly dialed down the seal drum to 50%. At first the vacuum got better and I was pleased. But then the vacuum got much worse, really quickly.

The vacuum tower pressure rose rapidly. And then things really went to hell. It transpired that there had been a small leak in the line connecting the vacuum tower heater to the tower. This is called the transfer line. It's normally under a slight negative pressure. But, when the vacuum broke, the leak pressured-up and vacuum tower feed at 730°F blew out into the atmosphere. The hot tar auto-ignited and a tremendous fire flared into being.

What had happened? I had uncovered a hole in the seal leg inside the seal drum by lowering the level (see Figure 13.2). Tail gas was sucked up through the hole. This had prevented the drainage of condensate down the seal leg. The condensate backed-up into the condenser and flooded over into the suction of the first stage ejector and the vacuum had broken. The operators had kept the level high in the condenser to keep this hole covered so that gas couldn't get sucked up the condensate drain line.

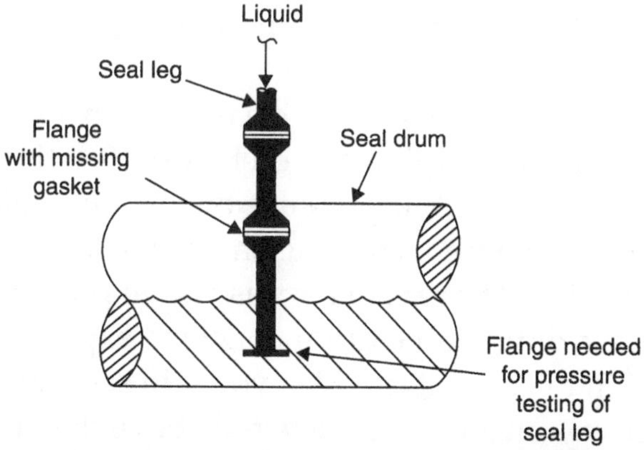

Figure 13.3 The seal leg inside the seal drum must be absolutely leak proof otherwise, the leg will not drain.

After the fire was extinguished, I apologized to Joe.

"It's okay, Norm. Everybody's gotta learn. But it's kinda better to learn by listening to us old operators rather than just that school book stuff."

It's a lesson I've long remembered.

Figure 13.3 shows an equally common problem. Note that the internal flange inside the seal drum does not have a gasket, hence the flange will certainly leak.

13.3 Seal Leg Flange Leak Outside Seal Drum

In the process industry, tiny problems may have giant consequences. For example, I was working on a very large steam turbine for Mobil Oil at their Coryton Refinery in England in 1986. The turbine was driving Mobil's very large fluid cracking unit air blower. High pressure motive steam exhausting to a vacuum surface condenser was employed to drive the huge turbine. The turbine was running slow because the condensing pressure in the surface condenser was high. The excessive surface condenser pressure reduced the amount of work that could be extracted from each pound of steam by about ten percent. As a consequence of this low turbine speed, the cracker feed rate was severely constrained by the lack of combustion air, due to the reduced air blower capacity.

The vacuum system itself was clearly not an issue. The operators were throttling (valve "A" on Figure 13.4) the inlet to the 1st stage jet to hold back pressure on the air cooled surface condenser, and hence increasing the exhaust steam pressure from the turbine. The reason for this was to promote drainage from the case drain of the turbine. Let me explain.

If you will refer to Figure 13.4, you will see that the surface condenser was an air cooler elevated above the turbine. If we employ a water cooler surface condenser, then the turbine is self-draining into the shell-side of the condenser.

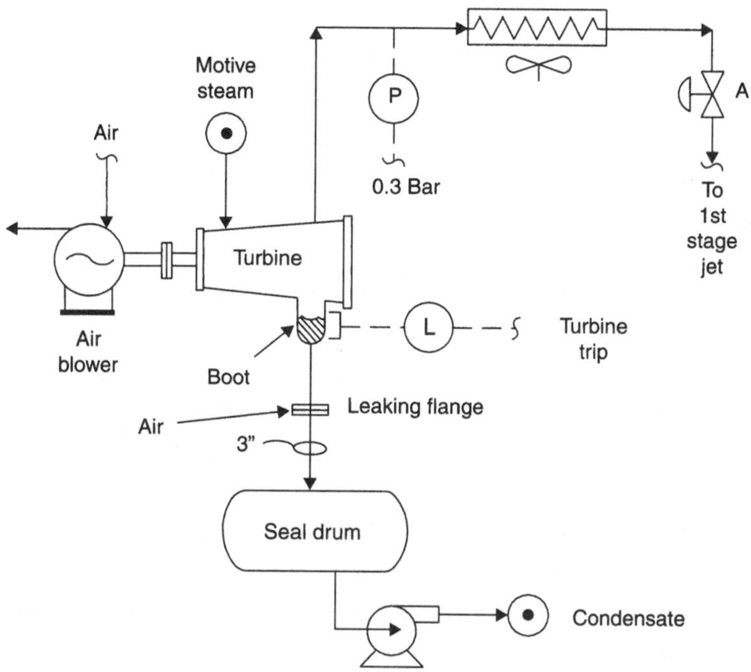

Figure 13.4 Leaking flange on drain line causes turbine to trip-off due to high water level in the boot.

But why do we need to drain the turbine case? Because, when we use steam – wet or dry – to drive a turbine, and exhaust the motive steam to a vacuum of 0.1 or 0.2 atmospheres absolute (i.e., 70 to 150 mm Hg) or less, perhaps 5% to 10% of the steam will condense as the enthalpy of the steam is converted to turbine shaft horsepower (an Isoentropic process). If this condensate accumulates in the boot of the turbine case shown in Figure 13.4, it will trip-off the motive steam to the turbine on high liquid level. This is the kind of occasion to consult your Mollier Diagram or steam tables.

The plant operators at Mobil, Coryton Refinery, had found that unless they held a back pressure on the air cooled surface condenser of about 0.3 atmospheres (220 mm Hg or 0.3 BAR absolute), they would get a high level alarm on the boot located at the end of the turbine. And this turbine exhaust back-pressure was robbing the air blower of horsepower.

The refinery management decided to shut the cracker down to clean the obviously plugged 3″ condensate drain line. This was washed out by a long water hose, but without disassembling the line. Upon start-up, the malfunction was found to be identical.

Referring to Figure 13.4 again, I'll restate the entire problem:

- The level in the small condensate pot at grade (seal drum) was visible in the level glass and correct.
- The level in the turbine boot was visible in its glass and correct.
- The drain line had worked properly for years, but now did not drain freely.
- The line had been cleaned and found to be free of obstruction.
- A significant pressure had to be held in the boot to promote drainage, by partly closing valve "A".

I thought about this problem for a few minutes. Perhaps the drain line was leaking. Air sucked into the line would have prevented the free drainage of condensate from the boot into the pot. If so, then the operators would have to hold some back-pressure on the boot, so that at the elevation of the hypothetical air leak, the sum of the:

- Drain line pressure plus the head of water at the leak
- Would equal atmospheric pressure

And would you believe it, Ladies and Gentlemen, at that very elevation that I calculated, there was a FLANGE.

I cleaned and dried the outside of the flange. Wrapped the flange with grey duct tape. My hands were shaking with excitement. I asked the operator to start to slowly open the back-pressure control valve (valve "A"). As he did so, the turbine boot level did not rise as before. But the air blower speed increased as the turbine exhaust pressure dropped to 50 mm Hg. This saved Mobil a huge amount of steam energy and restored the Cracking Unit Capacity.

I solved this problem very quickly because I had seen it all before. At the Good Hope Refinery, when I was tech service manager, I had found a leak on a seal leg on our vacuum tower 1st stage condenser drain (barometric) line to the seal drum, which I also repaired with duct tape.

13.4 Seal Leg Design

The vacuum-jet system – including the jets and condensers – will be designed by the jet vendor. The process engineer must design the seal legs, the seal drum, and determine the condenser elevations. Difficulty arises because the condensers are under a vacuum and cannot simply be drained to a sewer. There are two ways to approach this problem:

1. Vacuum pumping drums
2. Full elevation of the condensers

The simplest method is to locate the condensers high enough so that they will drain by gravity into the seal drum. To calculate this elevation, use:

$$H_c = 1.1\ (P_c)$$

Where

H_c = minimum height of the condenser bottom above the maximum liquid level in the seal pot, feet.
P_c = pressure in the condenser, inches of Hg

Figure 13.5 illustrates this arrangement. Note that the elevation calculated is not the total elevation of the condenser above

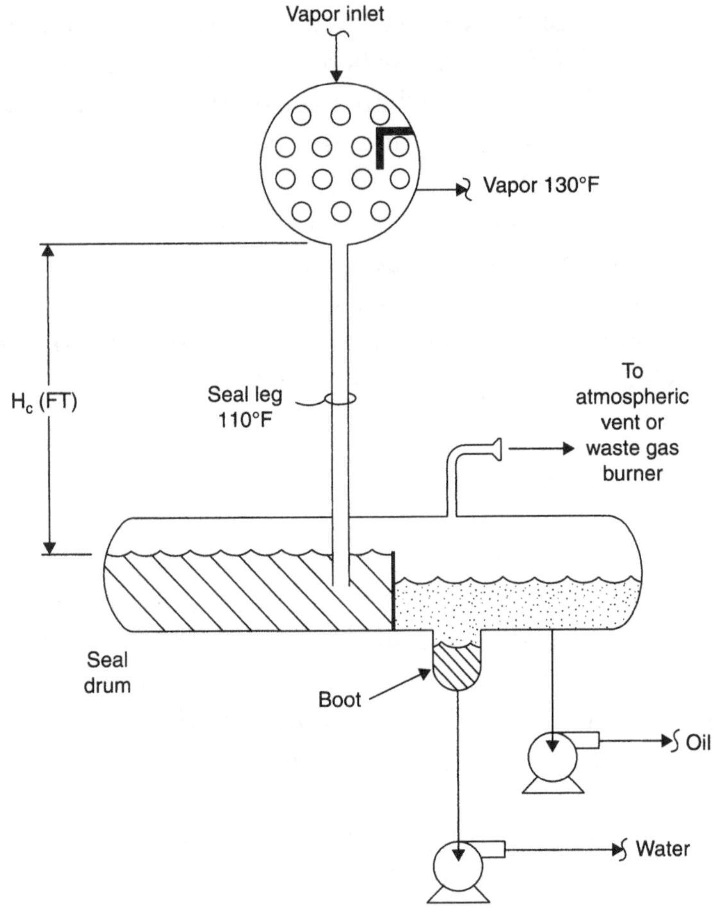

Figure 13.5 The vacuum condenser must be elevated above the seal drum for drainage.

grade. The seal drum itself will probably be 15 feet above grade. This will put the jet or pre-condenser 50 feet above grade. Each subsequent condenser can then be set at a progressively lower elevation. Of course, the other condensers will be much smaller than the pre-condenser (or first-stage condenser, if there is no pre-condenser), and so one might as well set all the jets and condensers on the same deck to simplify the piping.

In the "full-elevation condenser method" of sizing seal legs, the seal drum is vented to the atmosphere, and will run

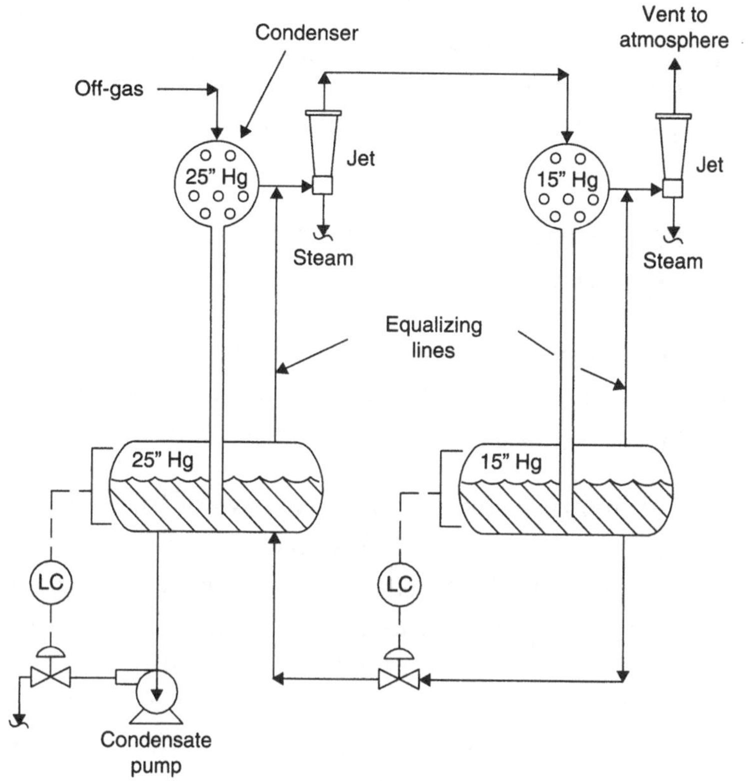

Figure 13.6 Maintaining the seal drums under a vacuum lowers the elevation of the condensers.

at a slight positive pressure. Only one seal drum is required, regardless of the number of condensers. On the other hand, with vacuum pumping drums as seal drums, a separate seal drum is required for each condenser, but the entire assembly can be located much closer to grade. Figure 13.6 illustrates this rather complex design. Each seal drum operates at the same pressure as the outlet of the condenser that it serves. This is achieved by means of the equalizing lines shown in Figure 13.6.

The principle disadvantage of this scheme, in addition to requiring multiple seal drums, is that the condensate pump must now be operated with a subatmospheric suction pressure.

This can be an operating headache because the condensate pump will cavitate when it has a leaking seal. Note in Figure 13.6 that the liquid from the second seal drum is pressured back to the first seal drum. This simplification saves one set of pumps.

13.5 Inadequate Seal Leg Length for Hydrocarbons

Sometimes barometric seal legs cannot drain properly even though there is no malfunction. Again, any back-up in condensate drainage from a condenser will result in:

- Submerging the bottom few rows of tubes which reduces the area available to condense the vapor.
- A consequent increase in the condenser vapor outlet temperature and flow to the downstream jet. A gradual loss in vacuum will result.
- Should the condenser level back-up to cover the bottom edge of the air or vapor baffle, then the vapor will tend to be trapped inside the condenser shell. A step change loss in vacuum is certain to result. That is, the operators will say that the vacuum "has broken."

For a condenser draining water (steam condensate) from a condenser operating at 38 mm Hg (i.e., 0.05 bar absolute of 0.73 psia), and a seal drum pressure of 3 psig, the absolute minimum elevation of the condenser boot above the seal drum is:

- $$\frac{(3\text{ psig} + 14.7\text{ psia} - 0.73\text{ psia})}{0.99\text{ S.G.}} \times 2.31 = 40\text{ ft.} \qquad (13.1)$$

- 3 psig = seal drum pressure
- 14.7 psia = atmospheric pressure in New Orleans
- 0.73 psia = condenser pressure

- 2.31 = feet of cold water (60°F) per one psi
- 0.99 = specific gravity of water at the 120°F in the condenser

At the Good Hope Refinery, at our new vacuum tower, we had no steam in the heater passes or in the bottom of the column. The overhead vapor product consisted of:

- Cracked gas
- Air leaks
- Condensible hydrocarbons boiling in the range of light kerosene or jet fuel

The kerosene all condensed out in the first condenser. Referring again to Figure 13.5, the design parameters were as noted above (equation 1), except that the specific gravity of 120°F kerosene was not 0.99 but 0.78. Thus the required minimum elevation difference (Hc) between the condenser and the liquid level in the seal drum was:

- $40\text{ ft} \cdot (0.99 \div 0.78) = 49\text{ ft.}$ (13.2)

But, the calculated elevation of the condenser was based on water, not kerosene. So that the condenser was 40 feet, not 49 feet above the seal drum liquid level.

The kerosene backed-up inside the condenser shell by about 12″–18″. I don't actually remember. But what I do remember is that the vacuum tower top pressure and the condenser pressure was a lot higher (perhaps by 100 mm Hg) than the design top pressure.

This all happened in 1980. I seem to remember we increased the elevation of the condenser to restore proper drainage. But after 32 years of girlfriend problems, children problems, financial problems, and hundreds of vacuum tower problems, I don't actually remember for sure what we did.

My daughter, Irene, who is typing this manuscript, called me about the above.

"Dad, that's nonsense. Look up your notes as to what actually happened."

"Irene. I'm writing this book from memory. I haven't kept a note in 47 years."

"From memory! Dad, you don't even remember what you had for breakfast!"

13.6 Inadequate Seal Leg Capacity

At the Good Hope Refinery, I reported directly to the owner, Jack Stanley. One evening Jack phoned my office. I was working late, trying to feel less depressed. It was Superbowl weekend in 1980.

"Norm, we just started up No. 2 Vacuum Unit. The vacuum is just awful. See what's up."

The problem was pretty obvious, as can be seen from Figure 13.7. The seal leg configuration is something I had never seen before and hope to never see again. At the time, our No. 2 vacuum tower was the largest in the world at 42 ft. I.D. The condensate drain from the second stage condenser was warm to the touch. The condensate drain from the first stage condenser was colder than the vapor outlet. Apparently, the first stage condenser was suffering from condensate back-up.

I suppose that combining the two condensate drains into one common seal leg exceeded the hydraulic capacity of the leg and that the lower pressure source of condensate was restricted. Regardless, I had a hose connected from a bleeder on the 1st stage condenser boot, to a connection below the overflow baffle on the seal drum, to allow this condenser to drain, without bucking the higher pressure flows from the #2 condenser. This solved the condensate back-up malfunction in the first stage condenser and allowed the No. 2 vacuum tower to develop its design vacuum.

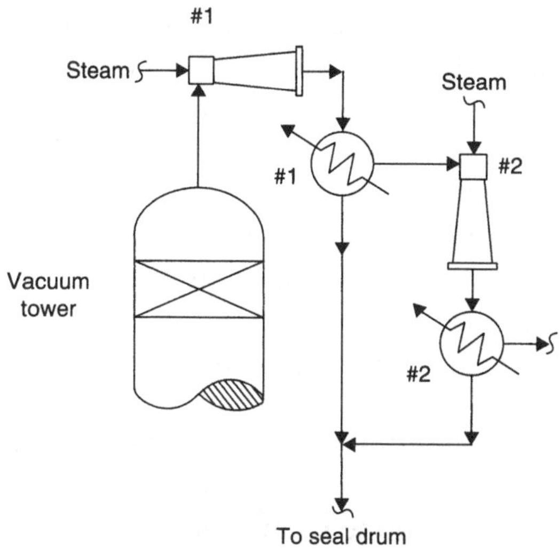

Figure 13.7 Connecting seal legs together results in a loss of vacuum.

A few nights later, Jack Stanley called again.

"Hey, Norm. I got an extra ticket to the Superbowl. Wanna go?"

And would you believe it, I turned him down. I never mentioned this to anybody before. Not even my wife or children. I thought I would have this story published after my demise. But now, compared with all my other dumb mistakes, it's no big deal.

13.7 High Back-Pressure from Seal Drum

At the Waste Oil Recovery plant, across the Mississippi River from our home in New Orleans, Texaco operated a small vacuum tower to recover clean gas oil from waste lube oils. Their ability to maximize recoveries was limited by inadequate vacuum.

My field pressure survey indicated that the problem was excessive 3rd stage ejector discharge pressure. This ejector discharged into a seal drum (see Figure 13.1 page 170) that vented to a waste gas burner in the vacuum heater. I measured the delta P across this burner at six psi. The design delta P, at the observed gas flow rate was two psi. Obviously the waste gas burner was plugged.

I've described how to measure the off-gas flow rate from the seal drum in Chapters 2 and 6 of this text. Basically, I use a measured flow of a tracer gas such as nitrogen and gas samples before and after the tracer gas is used to calculate the seal drum off-gas flow. A gas chromatograph analysis is required on each gas sample.

Waste gas burners can plug from a variety of problems:

- Sulfur formation – The waste gas will, at least in refinery applications, contain moisture, H_2S, and air. These components are certain to react to form elemental sulfur. Blackish corrosion deposits will make the sulfur appear to be coke, but if it melts or burns readily, it is not coke but sulfur.
- Gums – The cracked gases from the seal drum will contain diolefins (hydrocarbons with two double bonds). Upon heating to 300°F and exposure to air contamination from the seal drum tail gas, the diolefins form free radicals and polymerize. The resulting gums plug up the burners with corrosion deposits and the gums themselves.

Regardless of the mechanism, my experience teaches that the final condenser, seal drum and off-gas burner of a vacuum system are dirty services and require periodic cleaning. At the Texaco waste oil recovery facility, we temporarily bypassed the waste gas burner to their atmospheric vent. Cleaning the burner restored the vacuum to the original operating pressure.

13.8 Detecting Condensate Back-Up in Seal Legs

For a surface condenser, commonly used on steam turbine exhausts, the chances of restricting condensate drainage due to fouling are remote. But, in refinery vacuum tower operations, fouling deposits on the shell-side of surface condensers are certain. The cracked gasses combine with O_2 leaks to form gums. The HCl and H_2S, in an aqueous environment, form water insoluble $Fe(HS)_2$ corrosion products. The gums and iron sulfide particulates are the basis for shell-side fouling deposits and restrictive flow in the barometric legs.

If liquid starts to back-up in the surface condenser, you will observe that:

- The barometric leg gets cooler.
- The vapor outlet to the downstream jet gets warmer.

If condensate was draining freely from the condenser, the vapor and liquid outlets will be roughly the same. However, as we'll see in Chapter 15, "Air Baffle Problems," there is another problem with the surface condenser that causes the vapor outlet to be hotter than the liquid outlet. That malfunction is a defective air baffle seal to the shell. But I'll come back to this later on. For now, what I wish to describe is how to prove the condenser malfunction is condensate back-up.

The method is the opposite of what I tried in Sugar Creek in 1969. That is, raise the liquid level in the seal drum above the overflow baffle shown in Figure 13.2. If condensate is not already backed-up in the surface condenser, pushing up the liquid level in the barometric leg a few inches will not have any effect on vacuum or the surface condenser vapor outlet temperature. But, if vacuum is lost and if the surface condenser vapor outlet temperature does rise, then the condenser was already suffering from condensate back-up before you raised the level in the seal drum.

At the Amoco Refinery in Whiting, Indiana, the operators used to have to blow waxy deposits out of the seal legs with steam in the winter. They did this when the seal legs became cool to the touch. This requires a gate valve to be closed above the steam connection on the seal leg.

13.9 Condensate Back-Up Due to Air Leak in Barometric Drain Line

In 1998, I had an assignment at a refinery in Belgium. The problem was poor vacuum in the pre-condenser shown in Figure 13.8. The sketch and the story are in European metric units, rather than American units:

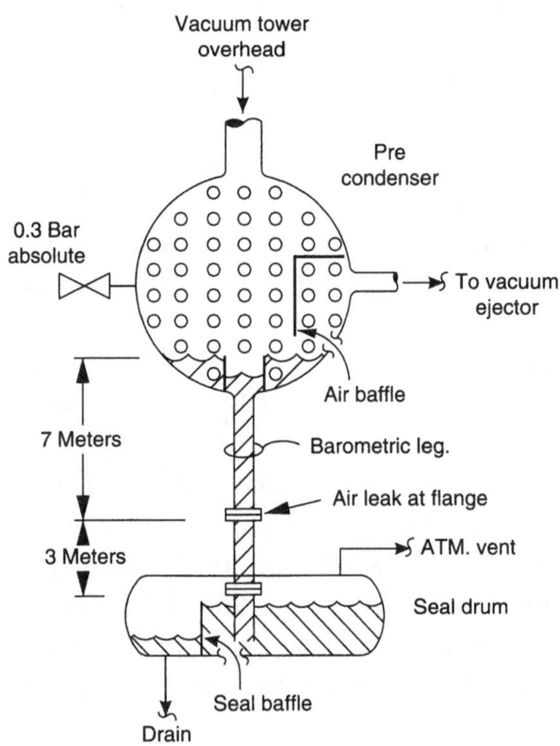

Figure 13.8 Air leak limits vacuum in the condenser. Pressure of flange balances with atmospheric pressure.

Table I One Atmosphere of Pressure.

European	American
10 meters of water	34 feet of water
One BAR absolute	14.67 psia
760 mm Hg	0.0″ Hg vacuum

Historically, the condenser had been able to develop a vacuum of 0.1 BAR (i.e., 76 mm Hg). Now, the best vacuum that could be developed was 0.3 BAR (i.e., 228 mm Hg). What was the malfunction that caused this limitation?

I noted a flange on the barometric drain leg, seven meters below the bottom of the condenser. Seven meters equals 0.7 BAR absolute. If I add the head of water equivalent to 0.7 BAR, to the 0.3 BAR absolute surface condenser pressure, I obtain 1.0 BAR absolute or atmospheric pressure (note that the refinery was located at sea level).

I cannot say that I observed that the flange was actually drawing air between the flange faces. However, I reasoned as follows:

- If the pressure in the surface condenser dropped from 0.3 BAR absolute to 0.2 BAR absolute, then the flange would have drawn air into the barometric leg.
- With air bubbling up through the leg, the pressure head of water required to promote condensate drainage would have been reduced.
- Water in the surface condenser would have backed-up and covered a portion of the tubes.
- Tubes submerged in water will not contribute to steam condensation, as the steam cannot contact the submerged tubes.
- The incremental uncondensed steam escapes from the side vapor outlet nozzle on the surface

condenser (see Figure 13.8). This increases the vapor load to the downstream vacuum ejector.

- The extra moles of steam flowing to the ejector will cause its inlet pressure to rise. This will increase the pressure in the surface condenser.
- The surface condenser pressure increases, so that at the elevation of the leaking flange on the barometric leg, the head pressure of water equals one bar absolute, or the pressure at sea level.
- This pressure stopped the air leaking into the flange, and restored drainage through the barometric leg. But any reduction in the surface condenser pressure below 0.3 BAR absolute, would reestablish the leak, until the system reached equilibrium.

Once again, I wrapped aluminum duct tape around the suspect flange. Having wrapped the flange, the vacuum in the surface condenser slid down to 0.1 BAR absolute. The interesting part of this incident, was that I could not actually observe that the flange was noticeably leaking. I drew the correct conclusion, based on my calculations, rather than on field observations.

Incidentally, I have presented this example many times in my seminars. When I explain the calculations using American feet and psi, no one understands. When I explain my calculations using meters and bars, everyone understands.

This story repeats the lesson I presented earlier in this chapter pertaining to the drain line from the steam turbine at the Mobil Coryton Refinery. So often, I encounter the same process problems, but in different services. And, this is such an example.

13.10 Seal Drum Design

Keep in mind that the vacuum system seal drum is a corrosive service prone to the accumulation of corrosion products and other fouling materials.

H_2S and air in an aqueous phase promotes biological corrosion. The cracked gas itself, when mixed with oxygen, will generate fouling deposits and gums.

The products of corrosion due to HCl in the vacuum tower overhead which corrode the surface condensers, will all be flushed into the seal drum.

The size of the seal legs and their internal clearances inside the seal drum should be generous.

A special note is required about the size of the seal drum hydrocarbon pumps. Don't forget that entrainment from the vacuum tower and poor naphtha recovery in the upstream crude tower can double or triple the required size of these pumps.

13.11 Seal Drum Fills with Corrosive Deposits

Jim Kane and I worked together during the five month strike at Texas City in 1980. Mostly, we worked 12 hour night shifts on the sulfur recovery plant. About 2:00 A.M., we ran around the unit five times, for a total of three miles. Jim was 25 and I was 38.

By 2010, Jim had gained 100 pounds and abandoned running. He had also become the plant manager of the Reading Refinery in Oklahoma.

"Norm, here's our problem. We run along at 15 to 20 mm Hg tower top pressure for a month or two. Then suddenly, with no warning, the vacuum breaks. Shoots up to 100 mm Hg. No warning. Just happens."

"But Jim, what initiates the incident?" I asked.

"Nothing. It just happens by itself," Jim answered.

"What do you all do to restore vacuum?"

"Norm, that's the strangest part. It just comes back by itself. Sometimes takes several hours, sometimes takes a day or two. The entire refinery is upset. Vacuum jumps off the chart. To over 100 mm Hg tower top pressure."

"Sounds bad Jim," I remarked.

"Yeah. Real bad. And getting more frequent. Had two incidents just last week."

Jim Kane glanced up at the clock. "Sorry, Norm. I've got to conference call the V.P. at 10:00 a.m. See what you can figure out. You can set up your computer in the tech service library. Let Bill Smit, the tech manager, know what data you'll need. See you for lunch."

I didn't have time to explain to Jim that I didn't eat lunch, or use computers, and had no use for data that came to me second hand. I just walked out into the refinery and started to explore the vacuum unit. But not for long.

"Hey you," challenged the unit outside operator. "What you doin' here? Who the hell are you?"

"I'm Norman Lieberman."

"Normal who? What you want?"

"I work for Mr. Kane. He asked me to figure out why the vacuum suddenly breaks and then comes back by itself."

"Man, that's the dumbest thing I ever heard. I don't know why the vacuum breaks. But it sure as hell don't come back by itself," explained the hostile young operator.

"Can I ask your name?" I inquired.

"Me? I'm Cal. Calvin Brown. The way we get that there vacuum to come back is to install a steam hose (see Figure 13.9) and blow the seal leg clear. Got to blow the crap out of the seal leg."

"Cal. How does that restore the good vacuum?"

"Look Libersky. That restores condensate drainage from the condenser. The backed-up condensate goes down and uncovers the tubes submerged by the condensate. Helps condense and cool the vapors. This unloads the downstream ejector and helps out the first stage ejector by lowering its discharge pressure," Cal explained.

"But why does it sometimes take a day or two, Cal, to restore the vacuum?"

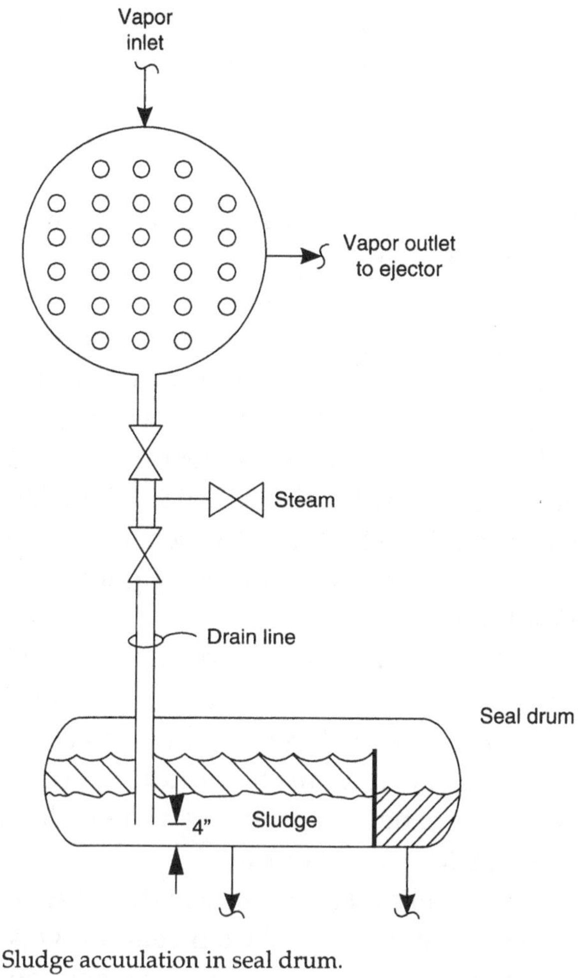

Figure 13.9 Sludge accuulation in seal drum.

"Cause, Mr. Letterman, the other shifts are too dumb or lazy to hook up the steam hose up there on the sixth landing. Only our "B" shift knows that trick."

"Calvin, I really appreciate your help," I said sincerely. "But I still don't understand what causes the loss of vacuum in the first place. What causes the seal leg to all of a sudden plug-off? And why is the problem getting worse?"

Calvin finally focused on my name stitched in gold on my blue coveralls.

"Mr. Lieberman. Let's get a cup of coffee and ask Ben Harris, the lead operator on our shift. Ben's real smart."

As Calvin and I sat drinking coffee in the break room with Ben, an old maintenance guy in faded gray coveralls, listened carefully.

"Excuse me, gentlemen," the ancient maintenance man interrupted, "Maybe I can help."

"What's up, Clyde?" Calvin asked.

"Well. During the turnaround last year, we opened up the seal drum. Found it full of sludge. Corrosion sludge, I reckon. Half full of solids. All them seal legs were buried in that sludge. Mr. Smit hired a chemical cleaning company to dissolve out all them deposits," Clyde said. "After the chemical cleaning was done, I peeked into the seal drum. Seemed like most of them deposits were still there. I guess that chemical didn't work all that good."

"Seems like it didn't, Clyde," I agreed. "Then what?"

"Well, you know this is just between me and you, Mr. Lieberman. And Cal, you don't need to spread this around neither. Mr. Kane, he told us to button-up the seal drum pronto. Said we were behind schedule and needed to get that vacuum tower back up and runnin'."

I checked the vessel sketch of the seal drum. As you can see from Figure 13.9, the bottom of each seal leg was only four inches above the bottom of the drum. Hence, only a small amount of sludge could result in condensate back-up. As I explained to Jim Kane, there is no reason for the seal leg to be more than a foot or so below the top of the overflow baffle. Therefore, we shut-down, cleared the deposits out of the vessel, cut 12 inches off of the seal leg and re-streamed the unit. The chronic lost vacuum incident disappeared and refinery stability was restored.

I still like to think about Jim Kane. Not the Jim Kane of 2010, plant manager, but rather my running partner of 1980. My slim, energetic pal during the long strike in Texas City. They seem like two different guys.

13.12 Seal Drum Design Tips

Figure 13.10 shows a properly designed seal drum.

- Any number of seal legs may be run to the same drum. However, seal legs from condensers, operating at different pressures must never be joined.
- The seal leg must be assembled inside the drum so that it will never leak. This is a rather vital point because once the seal drum is closed, there is no way that the seal leg can be inspected.

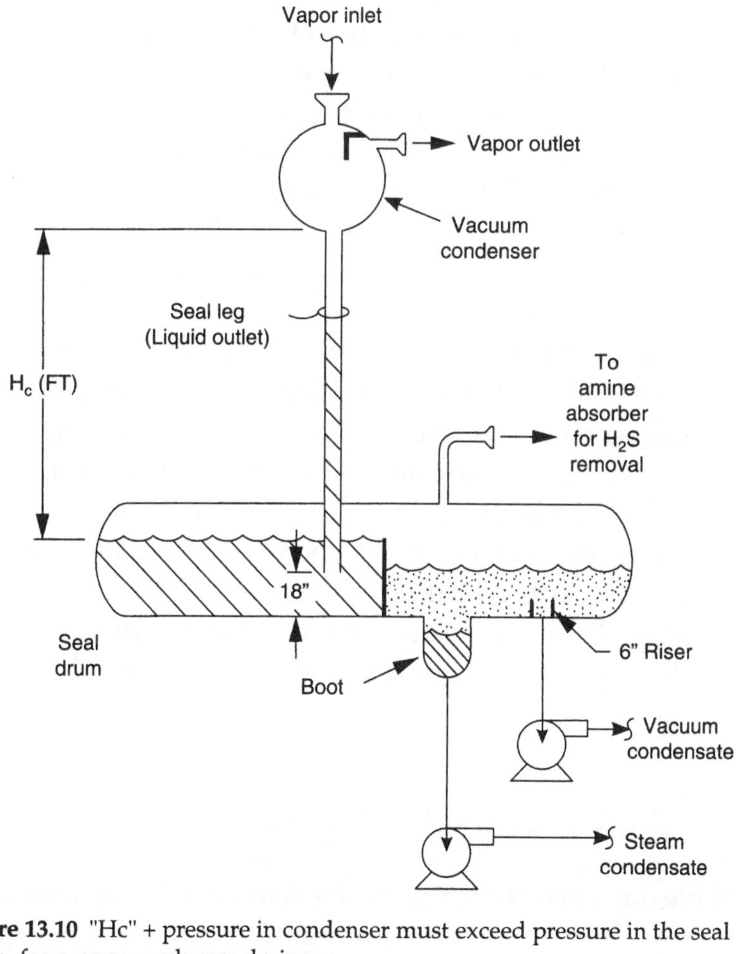

Figure 13.10 "Hc" + pressure in condenser must exceed pressure in the seal drum, for proper condenser drainage.

Once I was assigned to determine why a particular vacuum system could not pull its normal vacuum. After weeks of fruitless efforts, we opened the seal drum. There we found that no gaskets had been installed between the seal-leg internal flanges. As the seal drum was operating at atmospheric pressure, air was being sucked up into the seal leg and into the jet system through the leaky flanges. This situation illustrates the need for the bottom internal flange shown in Figure 13.3. Temporarily placing a blank flange over this internal flange allows the entire seal-leg assembly to be fully tested for leaks.

- The height of the seal baffle shown above the bottom of the seal leg is not too critical. A typical dimension for this height is 1–2 feet. The volume on the side of the seal drum that submerges the seal legs must be equal to or greater than the combined volume of the seal legs. In case the steam to the jets is accidentally shut off, this will prevent air from being sucked back up into the vacuum system.
- The liquid drained out of the condensers overflows the seal baffle into the oil-water separator side of the seal drum. This section of the drum is sized like any other oil-water separator—that is 7–10 minutes gross residence time based on net water production. A six inch riser is mandatory for the oil draw-off nozzle.
- The drum is vented to the atmosphere, but during normal operation no vapors should be created.

13.13 An Unfortunate Incident

For three days and three nights, we had been trying to start up the world's largest vacuum column. The huge tower, capable of recovering 110,000 barrels per day of gas oil from crude oil residue, was designed to operate at 29 inches of Hg vacuum. The best we could do was 22 inches of Hg.

The unit was equipped with two parallel trains of steam ejectors. As there was a steam shortage in the refinery, we were only using one of the two trains. This was not felt to be our problem because we could only pull 22 inches of Hg vacuum with absolutely no feed in the unit. Also, we knew that we were not overloading the single train of jets because of an air leak. We could block off the steam to the jets and still hold 20 inches of Hg vacuum for hours.

The problem was that both parallel condenser drains were tied into the same seal leg. Vapor from one condenser will recycle to the other parallel condenser backward through the idle jet and then back to the suction of the working jet (see Figure 13.11). This overloads the working jet. This problem only occurred when steam was cut off to one train of parallel jets.

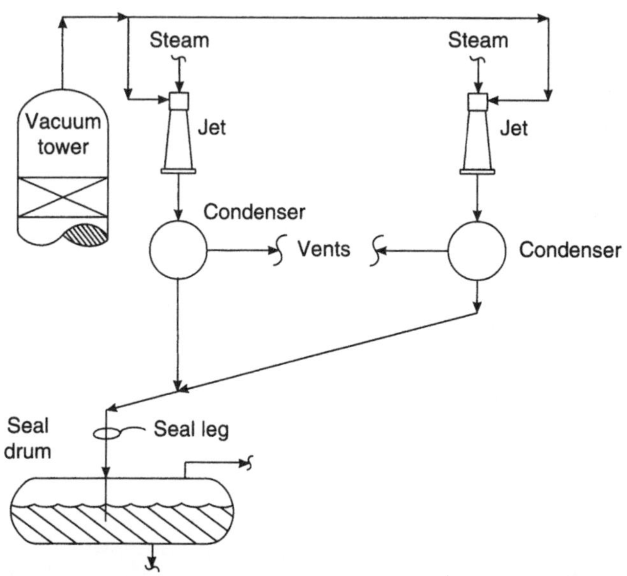

Figure 13.11 Connecting two condenser drains to the same seal leg was a design error which caused a loss in vacuum.

To rectify this problem, block valves were installed immediately below each condenser drain. When the steam to a jet was shut off, the drain line from the condenser, which served the idled jet, was closed off with the new block valve. The lesson is, do not tie seal legs together.

14

Other Types of Vacuum Equipment

14.1 Hogging Jets

Hogging Jets are used in several services:

- To extract air that leaks through the mechanical seals on shafts in condensing steam turbines.
- To air-free large vacuum systems rapidly to speed start-ups. Hogging jets are not intended for continuous services, even though they are habitually used for that purpose.
- To draw noxious vapors out of storage pits or tanks for disposal at a safer location. My experience is with the use of hogging jets on sulfur plant product sulfur pits. The H_2S that's dissolved in the sulfur flashes slowly out of the sulfur product in the underground concrete pits.

The hogging jet is just an ordinary steam jet, but lacks a converging section and thus never develops a sonic boost. It discharges directly to the atmosphere rather than through a condenser. Thus, there is no recovery of non-condensable gas and no seal drum. Other than poor motive steam conditions and steam nozzle wear, there isn't much that goes wrong with a hogging jet. I saw one working on a steam turbine surface condenser last month. It was discharging to the atmosphere with a suction pressure of about 150 mm of Hg. Atmospheric pressure was 720 mm Hg. The hogging jet was developing a compression ratio of about five to one. This is really good for such a jet.

On the other hand, the hogging jet is relatively inefficient, as compared to an ordinary converging-diverging jet, as far as motive steam consumption is concerned.

14.2 Use of Hogging Jet on Surface Condenser

I was working on a cracking unit (FCU) at the Coffeyville Refinery in Kansas. Coffeyville is a famous American town. In the 1870's, the Dalton Brothers tried to rob the two banks in the town. In the ensuing gun battle, four robbers and four of the townsfolk were killed. Since then, the town has receded back into obscurity, except for the Hollywood movie, "The Dalton Brothers Ride Again." (Which was not filmed in Coffeyville) The location of each dead body is commemorated by a flat bronze plaque, in the shape of a body, set into the sidewalks.

Coffeyville is located about 1,200 feet above sea level. Hence, atmospheric pressure is not 30" Hg (sea level), but about 29" Hg. I noted that the pressure in the turbine exhaust surface condenser was 23" Hg. This steam turbine drove the FCU wet gas compressor, which limited the unit's capacity.

The vacuum in the surface condenser was not being pulled by the regular two-stage jet system. This system, consisted, as it always does, of dual converging-diverging ejectors. This system had not been operated in years. Ordinarily it would develop vacuum of about 27″ Hg (i.e., 50 mm Hg. The 27″ Hg is 2″ Hg above a perfect vacuum of 29″ Hg in Coffeyville). There are 25 mm per inch.

The 23″ of Hg (i.e., 150 mm Hg. The 23″ Hg is 6″ Hg above a perfect vacuum of 29″ Hg in Coffeyville.)of vacuum was being pulled by the "Hogging Jet."

The hogging jet is a simple device. It works just the same as the diverging portion of a regular ejector, but it lacks the converging section. Typically, it develops a compression ratio of three to four. It uses a lot of steam compared to the converging-diverging jet. The hogging jet discharges to the atmosphere. At the Coffeyville Surface Condenser, the hogging jet was doing a terrific job. Its compression ratio was:

- (29) ÷ (6) = 4.83
- The 29″ Hg is atmospheric pressure.
- The 6″ Hg is the absolute pressure in the surface condenser.

However, even at an excellent 6″ Hg (absolute) vacuum for the hogging jet is pretty bad compared to the normal surface condenser pressure of 3″ Hg (absolute), for the expected vacuum developed by the regular two-stage ejector vacuum system. For the motive steam being used in the turbine, I calculated from my Mollier Diagram, that 10%–15% extra of motive steam was needed to drive the turbine, due to the higher surface condenser pressure of 3″ Hg (i.e., an extra 75 mm of Hg).

I asked the Coffeyville operators why they were using the hogging jet instead of their regular two-stage ejector system.

"Mr. Lieberman," the chief operator explained, "It's so much easier to use the little hogger rather than the regular complicated system."

"Yeah. That's true. But the higher condenser pressure causes you to increase steam flow to the turbine by 5,000 lbs/hr. That's like $450,000 a year worth of steam you're wasting."

"Really, Mr. Lieberman? Glad it ain't my money being wasted."

14.3 Liquid Seal Ring Compressors

An excellent alternative to pulling a vacuum with a steam jet is to use a liquid seal ring compressor. I have used these machines to pull a vacuum of 20 inches of Hg on a large vacuum tower. The compressor simultaneously developed enough discharge pressure to exhaust the vacuum tail gas through a burner and into a firebox.

Figure 14.1 illustrates the major components of a liquid seal ring compressor which is also called a vacuum pump. The pump case is half-filled with water or diesel oil. The liquid is thrown to the periphery of the casing and forms a liquid ring which seals the space between the impeller blades and the casing. At the top position of the impeller, the chambers between the impeller blades are filled with liquid. As the eccentric impeller rotates, the liquid ring moves away from the hub, increasing the space in the pumping chamber. This draws gas into the chamber from the inlet port next to the impeller. As the impeller revolves, gas in the impeller chambers is compressed by the liquid ring and is expelled through the outlet port.

The liquid inside the case also absorbs the heat of condensation and compression. Hence, liquid must be continuously circulated through the case by means of an external pump. The circulating liquid must be cooled before returning to the compressor.

The operating costs, in terms of energy requirements, for liquid seal ring compressors are a small fraction of the steam consumed in jet ejectors. However, these vacuum compressors usually require a great deal more maintenance than do steam

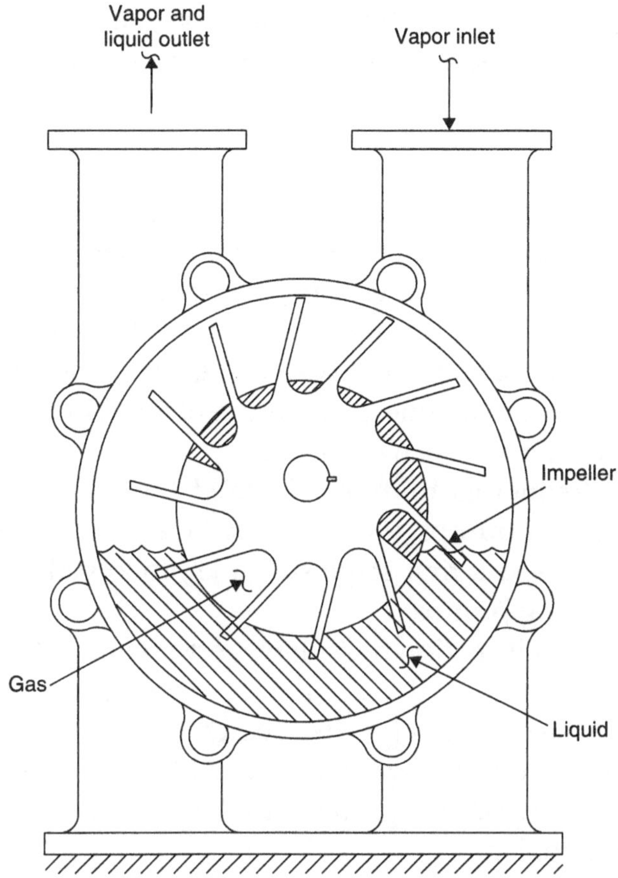

Figure 14.1 Liquid seal-ring vacuum pump. Mechanically reliable, energy efficient, but expensive compaired to steam ejectors.

jets. Also, they are relatively tricky to operate. For instance, the rate of liquid circulation through the compressor case must be precisely set. Too little liquid flow will lead to low compressor efficiency; too much flow can cause the eccentric impeller to bog down and result in equipment damage.

On balance, a liquid seal ring compressor should be used in preference to a steam-jet ejector when the following conditions prevail:

- The compressor is located in an area where it can be frequently checked by operations personnel.

- In large applications, which ordinarily require several thousand pounds per hour of steam, if jets are used.
- Steam for vacuum jets is costly or in short supply.

Do not specify liquid seal ring vacuum compressors to save energy and then include a backup steam jet in the design. Whenever I have seen this done, the operators used the zero-maintenance steam jet and abandoned the compressor as requiring excessive maintenance. Use the same philosophy for a liquid ring machine as for any centrifugal pump in critical service – provide a 100% spare.

Texaco Refineries routinely favored the use of two-stage liquid seal ring pumps, which worked just fine. For deep vacuums, they used a first stage steam jet, followed by two, in-series, seal ring pumps.

The only malfunction that I have experienced with the seal ring compressors at the Good Hope Refinery was organic fouling of the circulating water cooler. H_2S consuming bacteria accumulated in the water circulated through the compressor case. We cured this problem by biocide injection into the inlet to the cooler.

From the designer's perspective, these vacuum pumps are more energy efficient than vacuum jets, but also more costly to purchase and install. Also, a small continuous water make-up stream prevents loss of level in the compressor case.

14.4 Gas Ejectors

In 1984, I was working in the gas fields of Laredo, Texas. I designed a recovery system for low pressure natural gas from a depleted well using a natural gas eductor. The motive gas for the jet was not steam, but high pressure natural gas from a second well. The high pressure gas was flowing thru a wellhead choke to reduce its pressure down to the collection header pipe pressure. So, instead of wasting this source of energy, I replaced the choke with a small eductor which boosted the pressure of the smaller gas flow from the low pressure well. This saved the rental and operating cost of the small wellhead compressor that would ordinarily be used for this job.

In 1965, I was working at a large crude distillation unit in Whiting, Indiana. The vacuum tower off-gas from the seal drum (at one psig) was compressed up to about ten psig into a compressor suction drum, using a small hydrocarbon eductor. The motive gas was the crude unit wet gas at about 30 psig, which also had to flow into the same compressor suction drum. I don't actually recall the relative motive gas to recovered gas flow in either of the above cases. But, in general, the motive gas flow would be about three times greater than the recovered gas flow. In the second case, the motive steam for the eductor was saved. In both cases, the ejector (eductor) operation was trouble-free and ran without any operator intervention.

14.5 Liquid Ejectors

My experience with liquid ejectors is limited to one large system in Lithuania, built by the Russians in the 1980s. The vacuum tower had a pre-condenser and the pressure in the pre-condenser closely corresponded to the vapor pressure of water at the pre-condenser vapor outlet temperature. Thus, the two-stage liquid ejector system was never an issue, and was working properly. Diesel circulated through a cooler and back to the ejector system. The fact that no one in the plant knew much about the system is ordinarily a good indication as to reliability. My lack of experience with this sort of vacuum system does not imply that it is in anyway inferior to steam ejectors.

14.6 Ejector Compression Efficiency

I have stated that a steam ejector is like a two-stage compressor with no moving parts. Mechanically, it is robust and simple compared to other types of compressors. Certainly it's purchase cost and installation is just a small fraction of either reciprocating compressors or centrifugal compressors. Furthermore, as I have explained in this chapter, it can, and I have, used ejectors in non-vacuum services.

Why then, are steam ejectors not in as wide spread use as gas compressors?

The problem is compression efficiency. For example, a reciprocating compressor will have a compression efficiency of 80% to 90%. By efficiency, I mean the conversion of shaft horsepower to compression work.

A centrifugal compressor will have a compression efficiency of 65% to 80%. While centrifugal efficiency is less than reciprocating, the centrifugal compressor is more mechanically reliable than a reciprocating compressor.

The comparable compression efficiency of a steam ejector is about 25%, assuming it's a converging-diverging ejector developing its sonic boost (i.e., in its critical mode of operation).

After all, my clients sometimes replace steam ejectors with liquid ring seal pumps, as described in this chapter, on the basis of energy (i.e., steam savings) conservation.

In summary, steam jets are used because they are cheap to install, look simple to operate compared to a reciprocating machine or a rotating type compressor, but they waste motive steam.

The problem is that in reality, steam ejectors are not simple to operate at all. Likely, my clients would be better off (as Texaco did) using two liquid ring seal pumps in series, and not get involved with steam jets in vacuum towers. But then, I would lose 25% of my business and nobody would buy this book.

15

Air Baffle and Impingement Plate in Surface Condensers

The air or vapor baffle inside the surface condenser is critical to proper performance of the downstream vacuum ejector or vacuum liquid ring seal pump. Figures 15.1 and 15.2 illustrate the function of the air or vapor baffle and seal strip. The inlet vapor flows down across the tubes and underneath and back up inside the air baffle. The objective being to minimize the temperature and the moisture content of the residual vapor flow to the downstream vacuum jet. The usual malfunction is leakage of hot vapor between the edge of the air baffle and the shell I.D. The cause of this malfunction is defective seal strips that ought to seal the baffle, which is part of the tube bundle, to the shell I.D. The causes of the malfunction to the seal strips are corrosion or improper installation of the bundle. The leaking seal strips permit the hot, moisture laden vapor, to bypass the tube bundle and blow directly into the downstream jet. This overloads the jet, and diminishes vacuum.

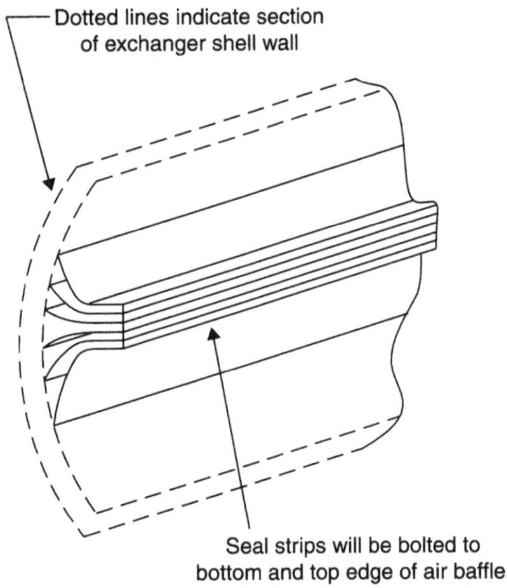

Figure 15.1 Detail illustrating how a modern air or vapor baffle seals up agaist the inside of the exchanger shell. Seal strips are stainless steel, not brass or copper.

It's simple to identify that the air baffle seal strips have failed. Check the skin temperatures along the shell of the surface condenser at the elevation an inch or so below the seal strips with your infrared gun.

At the one refinery, I reduced the vacuum tower flash zone pressure from 75 mm Hg to 62 mm Hg, by renewing the seal strips on the second stage condensers. I've discussed this incident later in this chapter.

15.1 Mechanical Configuration of Seal Strips

Figure 15.1 shows a drawing of the seal strips attached to the top edge of the air or vapor baffle. The purpose of these seal strips is to prevent hot vapor in the feed from bypassing 90% of the tubes in the condenser, and overloading the downstream ejector with hot gas and steam. This sketch

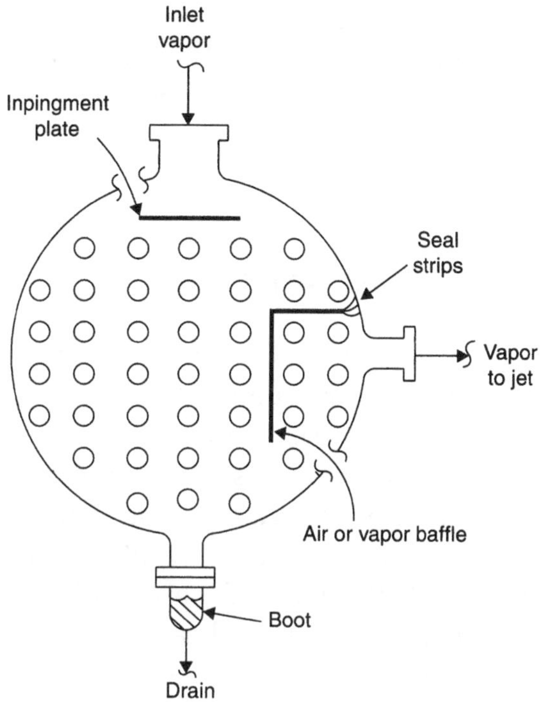

Figure 15.2 Internal components of a surface condenser Seal strips retard leakage around air baffle.

shows three pairs of seal strips. More commonly, there are four or six pairs. The seal strips are clamped to the edge of the air baffle. Each seal strip is perhaps three or four inches wide with about the thickness of a beer can. When purchased, the material comes in a roll that can be cut to length. The appropriate length is the space between the two tubesheets of the condenser bundle, less an inch or so.

The method of attachment of the seal strips (i.e., clamped to the air baffle) should suggest to my clients that they should be replaced each time the bundle is pulled. The tube bundle cannot be extracted for cleaning without damaging the seal strips. I used to think that replacement of seal strips required supervision by the condenser manufacturer. Now that I've seen the construction of the vapor baffle in the field, I realize it's something I could do myself without expert direction. For refinery applications, copper based alloys should never be used for seal strips. My material of choice is 316 (L) stainless steel.

15.2 Corroded Brass Seal Strips

One of the main causes of leaks around air or vapor baffles, is corroded seal strips. This happens because the seal strips are constructed out of copper based (brass) alloys.

Most surface condensers are used in steam turbine exhaust service, where copper or bronze type seal strips are the material of choice. However, in refinery applications, the off-gas contains NH_3, HCl, H_2S, HCN, CO_2, and moisture. These corrosive components, when wet, will cause the brass seal strips to break off. In one condenser, the vapor outlet temperature was 30°F hotter than the liquid outlet, due to leakage around the internal air baffle. We replaced the copper type seal strips with stainless steel strips during the next turnaround, which eliminated this condenser malfunction.

15.3 Air or Vapor Baffle Leak

I've observed this problem both at the Valero Refinery in Aruba and the Texaco Port Arthur Refinery, as well as a half dozen other plants.

Figure 15.2 is a cross-section of a pre-condenser or a surface condenser. Figure 15.3 is the side view of this sort of condenser. The cross-section shows the edge view of the vapor or air baffle. The purpose of the air baffle is to prevent the hot inlet vapor from bypassing the tubes and blowing out of the outlet nozzle. The hot vapor will contain many moles of steam, which will overload the downstream jet.

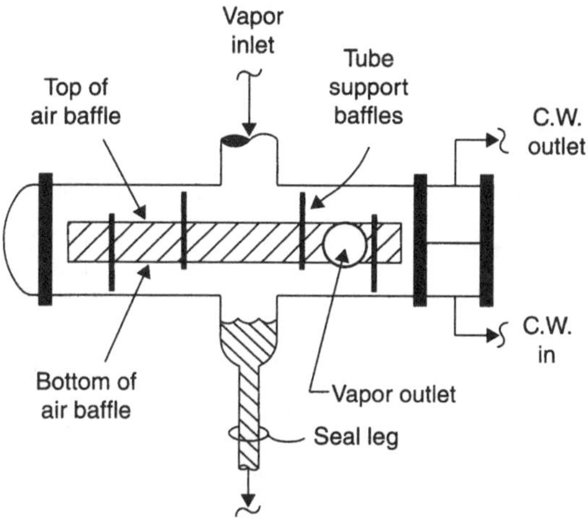

Figure 15.3 Side view of air baffle. Tubes and floating head not shown.

The purpose of the seal strips, shown in Figure 15.2, are to seal off the opening between the top edge of the air baffle and the shell I.D. These seal strips are clamped onto the air baffle. They are used in three to five pairs. Typically, each strip is around 3″ wide, perhaps 0.04″ thick and the full length of the vapor baffle. If used in a refinery vacuum tower, the seal strip metallurgy is 316 (s.s.) or 317 (s.s.). If used in steam turbine surface condensers, copper alloys (brass) are employed.

At the Texaco plant in Port Arthur, brass was accidentally used in a refinery vacuum tower surface condenser, and the copper alloy strips were severely damaged by corrosive H_2S, HCl, and NH_3. At the Coastal Refinery in Corpus Christi, I recall pulling bits of copper alloy seal strips out of the pre-condenser's boot pump suction strainer.

15.4 Identifying Defective Seal Strips

Figure 15.4 shows the skin temperature pattern I observed on the outside of the shell on one of three second stage

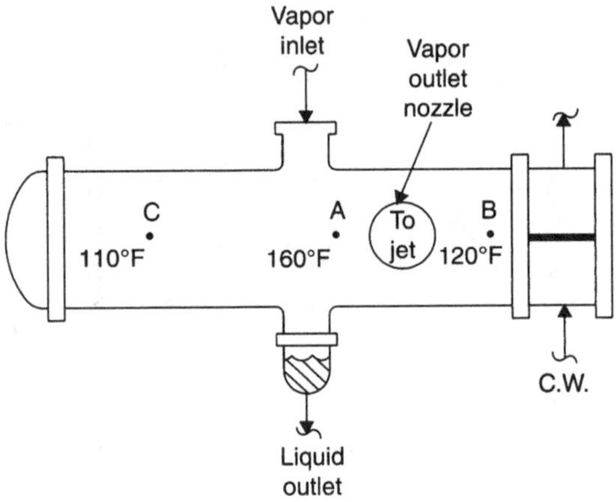

Figure 15.4 Skin temperatures on shell surface indicate an air baffle leak inside the surface condenser.

condensers for the Aruba vacuum tower. The hotter shell skin temperatures along the middle of the air baffle, and the colder skin temperatures along the outer portions of the air baffle, are a sure sign that the seal strips are leaking. Hot inlet vapor is preferentially blowing directly into the second stage jet, rather than circulating through the condenser bundles.

I explained my observations to Mr. English, the plant manager in Aruba. He decided to renew the seal strips on all three parallel sets of vacuum ejectors and condensers.

As a result, the inlet temperature to the second stage ejectors dropped from the 165°F, to an average of about 125°F to 130°F. Unloading the second stage ejectors reduced both the discharge and suction pressure of the first stage jets and thus, the vacuum tower top pressure.

It isn't that I don't recall these intermediate pressures; I just never bothered to measure them in the field. But what I really do remember is Mr. English phoning me at my home in New Orleans.

"Hey, Lieberman, it's English here."

"Yes, sir," I responded.

"Lieberman, the vacuum tower flash zone pressure has dropped from 27½ inches of mercury (i.e., 62 mm of Hg) to 28 inches of mercury (i.e., 51 mm of Hg). Lieberman, that's the best damn thing you've ever done at the Aruban Refinery. Great work."

Note that in Aruba, which is at sea level, full vacuum is 30 inches of mercury, or zero mm of Hg. In reducing the flash zone pressure from 27½" to 28" of mercury, I had reduced the flash zone pressure by around twenty percent. No wonder the plant manager was so pleased. This would certainly increase recovery of the heavy gas oil from vacuum tower bottom's tar by a very large quantity.

15.5 Air Baffle Clearance

One of the problems that I've seen recently on a surface condenser at a very large Gulf Coast refinery vacuum tower relates to the dimensions of the air baffle. In this case, the delta P thru the surface condenser was 27 mm Hg (i.e., 97 mm Hg inlet to 70 mm Hg outlet) at 50% of the design capacity.

The bottom edge of the air baffle was extremely close to the bottom of the shell. In theory, if the exchanger was clean, the open area between the bottom few rows of tubes would have been adequate for the non-condensibles to escape into the vapor outlet nozzle at the top of the air baffle (see Figure 15.2). However, as this is a rather serious fouling service (described below), the fouling deposits, which accumulated at the bottom of the shell, tended to choke off the vapor flow and create a highly localized pressure drop. As a minimum then, I would tend to terminate the bottom of the air baffle about 25% of the shell I.D. above the bottom of the shell.

The above criteria do not apply to turbine exhaust surface condensers which are inherently clean.

15.6 Fouling Mechanism in Vacuum Tower Surface Condensers

The fouling material that I typically observe in the pre-condenser and/or first stage condenser is a black sticky material. Further downstream (in the 3rd stage condenser), the black semi-liquid deposits become more gritty and greasy. The material in the pre-condenser appears to be polymerized hydrocarbons due to the presence of air and di-olefins, plus iron sulfides and nitrogen (filming and neutralizing amines). Lab tests showed at one plant that the pre-condenser deposits were quite soluble in methanol, but I imagine that they would be somewhat soluble in heavier alcohols that would not vaporize under vacuum conditions (i.e., Isopropanol). The deposits were, of course, totally insoluble in water. Otherwise they would have been dissolved in the condensing steam. Yet, in spite of this obvious lack of solubility in water, my clients try to clean surface condensers with hot water and then are surprised when they do not do any good.

15.7 Surface Condenser Impingement Plate

To protect the tubes from the erosive velocity of the shell-side inlet vapor, an impingement plate is used, as shown in Figure 15.2. The upper rows of tubes must be removed to make room for this impingement plate. This is both good and bad. It's good in that the open area created along the top of the shell allows the vapor feed to distribute laterally across the entire length of the tube bundle. It's bad in that the impingement plate, if too large, will restrict the entry of the vapor feed into the tube bundle. At the Texaco plant (now Delaware City Refining), I revamped their vacuum tower pre-condenser by reducing the condenser's impingement plate area by ⅔ with a consequent reduction in delta P from about 20 mm Hg, to about 10 mm Hg. A 10 mm Hg delta P is still excessive.

A far more dramatic change, but in the wrong direction, was achieved at the Huntway (now Valero) asphalt plant near San Francisco. I had completed a simple revamp design of this 14,000 BSD facility. I had measured the delta P across the pre-condenser at 5 mm of Hg before the unit was shut down for their turn-around. Guy Young, the plant manager, suggested that I include in the scope of my design, re-tubing the pre-condenser tube bundle.

"Actually, Guy," I responded, "The pre-condenser is working okay. The delta P is normal and its temperature approach is fine."

"Norm," Guy asked, "What do you mean by the temperature approach?"

"I'm looking at the condenser vapor outlet temperature minus the cooling water inlet temperature. A real good temperature approach is 6°F to 8°F. A real bad temperature approach is over 20°F. Your approach is 7°F. That's just fine."

"Well, we've had tube leaks in the past. I'll just play it safe and have the bundle re-tubed. We've got the original exchanger drawings. I'll just send the bundle off with the drawings to the local fab shop to be re-tubed," Guy decided.

Two months later, Guy Young phoned me at my home in New Orleans.

"Lieberman. You've ruined our vacuum tower with your revamp design. We can't make asphalt anymore. The flash zone pressure has doubled and the tower top pressure has tripled. You'd better get your skinny ass on a plane pronto."

So I got on a plane and flew to San Francisco the next day. The first measurement I made at the Huntway plant the next day was the delta P across the pre-condenser. Its delta P had increased from five to fifty mm of Hg.

"Guy," I explained, "Something has gone wrong with the pre-condenser. They screwed up the bundle when they re-tubed it. The delta P is ten times larger than before."

"Okay, Norm. Maybe it's not your fault," Guy said. "Drive over to the shop in Berkley. Maybe they've done something wrong."

But Mr. Kolwaski, the fab shop manager was most indignant when I suggested his workers had misassembled the condenser bundle.

"Look here, Mr. Lieberman. We've assembled that bundle exactly as per the drawings provided by Huntway. Even the impingement plate. That old bundle Huntway sent us didn't even have an impingement plate. Totally missing. I had to make a new one. Didn't even charge Guy Young for it."

What had happened? Many years ago, long before Guy Young became the plant manager, some smart engineer had the impingement plate removed to reduce the pre-condenser delta P. While that may have contributed to the tube leaks, it certainly reduced the shell-side pressure drop a lot. Mr. Kolwaski, consciously working from the original bundle drawings, had accidentally reintroduced the excessive condenser pressure drop by reinstalling the original over-large impingement plate.

15.8 Oversized Impingement Plate

At the Marco Refinery in Beaver City, I made a significant improvement in vacuum by reducing the delta P in the first stage condenser. This was done by reducing the size of the impingement plate, shown in Figure 15.2. The impingement plate was far larger than the inlet nozzle diameter, which introduced unnecessary shell-side pressure drop. A delta P measured on a condenser in vacuum service above six or eight mm Hg, should be considered a sign of a malfunction, either due to fouling or a flawed mechanical design.

Incidentally, this is the same vacuum tower that was suffering from the mal-functions I discussed in Chapter 3 of this text.

15.9 Impingement Plates as Vapor Distributors

Using an impingement plate below the vapor inlet of a surface condenser is intended to protect the upper row of tubes from erosion. Caution is advised! As described in the preceding story at the Huntway Asphalt plant in San Francisco, my client installed a new tube bundle with too large an impingement plate. The pre-condenser delta P increased from five mm Hg to fifty mm Hg, and asphalt product specs could no longer be met.

However, an extra large impingement plate, if designed properly, can improve the overall heat transfer efficiency of a surface condenser or pre-condenser without excessive delta P. Let's assume that the bundle is twenty feet long. Also, that

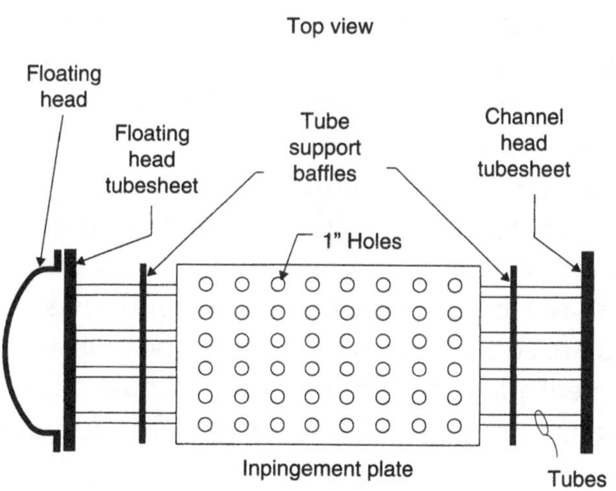

Figure 15.5 Inpingement plate in a surface condenser can promote improved vapor distribution. Total hole area equals nozzle inlet area. You are loooking down on the top of the tube bundle. Inlet nozzle size was eight inches.

it has four tube support baffles, a floating head, and channel head tubesheets. The vapor inlet nozzle is presumed to be in the center of the shell as shown in Figure 15.3. The impingement plate should then be designed as follows, to promote vapor distribution:

- Maximize the width consistent with the shell dimensions at the upper row of tubes.
- The length of the impingement plate should extend between the furthest tube support baffles. This will cover 60%–70% of the horizontal distance between the two tube sheets.
- Drill one inch, or one and a half inch holes, equally spaced, through the impingement plate, including the area beneath the vapor inlet nozzle.
- The sum of the area of these holes should be equal, or slightly greater, than the area of the vapor inlet nozzle.

I've copied this design from exchangers I've seen in a similiar service. I can't testify as to how much benefit is derived from such an impingement plate vapor distributor. Incidentally, if your condenser has the shell inlet located at the opposite end of the exchanger from the vapor outlet, this concept does not apply, and a conventional impingement plate is required to protect the tubes from erosion.

I've provided a not-to-scale sketch of this idea in Figure 15.5.

Graham, the leading manufacturer of vacuum equipment, no longer uses impingement plates. Several rows of, "dummy tubes," are used that have the same purpose as my perforated impingement plate design, shown in Figure 15.5. I consider such impingement tubes to be a reasonable design practice.

16

Optimizing Vacuum Tower Operation

I was all set to land a nice contract to revamp a vacuum column at the Unocal Refinery. The revamp of the tower was intended to recover an incremental 3,000 BSD of gas oil from the vacuum tower bottoms.

The process conditions shown in Figure 16.1 were the design-basis operating parameters. Note that the 715°F flash zone temperature and the 25 inch Hg flash zone pressure (128 mm Hg) are indicative of an operation that results in excessive gas oil left in the vacuum tower bottoms. This downgrades virgin gas oil from FCCU feedstock to delayed coker feed at a penalty of $20/bbl. A properly designed and operated vacuum column that employs steam in the heater coils operates at 27 inches Hg flash zone pressure, 760°F flash zone temperature, and 780°F heater outlet.

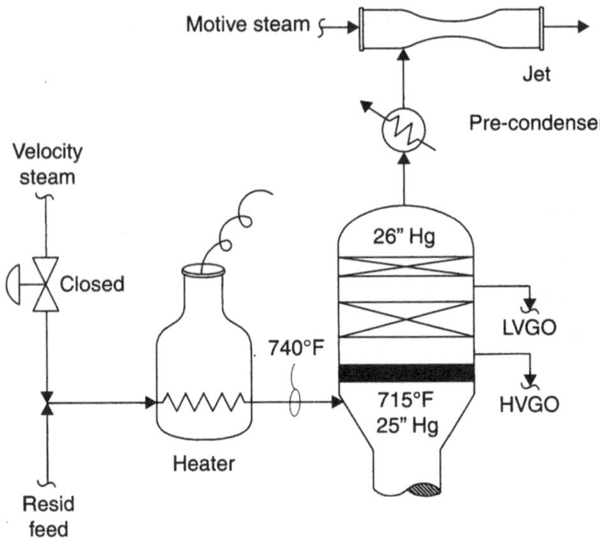

Figure 16.1 Velocity steam improves flash-zone pressure by surpressing thermal cracking in heater.

16.1 Steam to Heater Passes

However, there was no steam going to the heater coils. The operating superintendent said that he was not told to use steam in the heater passes. He asked about the reason for steam in the heater coils. I told him the steam is used to suppress temperature peaking in the heater. In vacuum heaters (and to a lesser extent, delayed coker heaters), the maximum coil temperature does not occur at the heater outlet because most of the absorbed heat goes into heat of vaporization rather than into sensible heat. Most of the vaporization occurs in the last 5 or 10 tubes of the heater.

Figure 16.2 illustrates the effect. Even though the absorbed heat is similar in each tube, the temperature of the vacuum tower charge drops due to the rapid vaporization of the resid as it approaches the heater outlet.

When steam is added to the heater passes, earlier and more uniform vaporization of the resid is promoted. The steam,

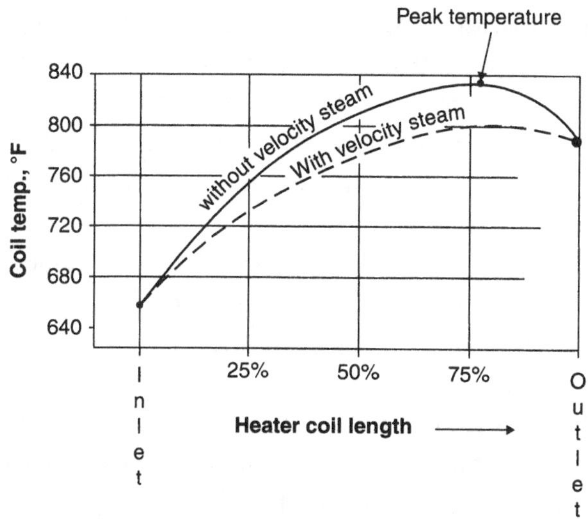

Figure 16.2 Velocity steam surppresses peak temperatures and evolution of cracked gas. The 800°F outlet shown is absolute maximum in my experience.

therefore, suppresses the peak temperature reached in the heater passes.

The production of cracked gas in the heater is largely a function of the peak temperature developed inside the heater coils. When the peak temperature is suppressed, the load of cracked gas to the vacuum tower overhead steam ejectors is reduced. The ejectors can, therefore, pull a deeper vacuum, lowering the tower flash zone pressure and increasing gas oil recovery.

The amount of steam can be determined by adding enough steam or condensate so that the pressure drop through the heater increases by 50% to 60%. Over a number of days, several adjustments were made to the tower operating parameters.

As a first step, steam was supplied to the heater passes on the unit and the flash zone pressure dropped from 25 inches Hg to 27 inches Hg (i.e., vacuum improved by 50 mm Hg). Gas oil recovery increased by 1,800 BSD.

The heater outlet temperature was then raised by 10°F. The superintendent questioned this move because in the past, raising the heater temperature overloaded the ejectors with cracked gas. But, because of the effect of coil or pass steam, the heater passes were running cooler even though the heater outlet temperature was higher.

The increase in heater outlet temperature raised the flash zone temperature to 723°F from the previous 715°F. Pressure rose a bit to 26.9 inches Hg, but the gas oil production rose another 800 BSD.

Then, steam pressure to the ejectors was adjusted to achieve the highest possible vacuum at the top of the vacuum tower. Cooling water pump turbine speeds were increased to push the maximum flow of cooling water to the overhead condensers. Also, the condensers were back-flushed.

After these moves, the flash zone pressure decreased, and the heater temperature was increased another 5°F. Finally, the vacuum tower bottoms stripping steam rate, heater outlet temperature, and heater pass flow were all optimized so as to obtain the maximum gas oil production. These last items usually take some trial and error.

The final result on this tower was a 3,200 BSD gain in FCCU feed at the expense of the delayed coker feed. This was about 200 BSD more than the revamp was supposed to achieve. At $20/bbl, that amounted to a $64,000 per day yield improvement on the tower.

It turned out that after the operating adjustments on the tower were made, there was no need for revamping the tower. Unfortunately, I never did get a contract for the revamp.

16.2 LVGO Pan Level Loss Causes a Loss in Vacuum

Oscar Wyatt, the man I admire most in the world, the founder of the Coastal Corporation, has recently been released from a federal prison. His "crime" was buying crude from Iraq illegally after the first Gulf War and befriending Saddam Hussein.

One of Mr. Wyatt's many accomplishments was the constructing of the Pelican Oil Refinery in Lake Charles, Louisiana. Perhaps I'm exaggerating in calling this a refinery. Its capacity was only about 5,000 BSD. The only finished product produced was paving asphalt for local sales to the Louisiana Highway Department.

The vacuum tower overhead system consisted of a very efficient liquid seal ring pump. I call this a pump, but it's really a rotary positive displacement type of a compressor (see Chapter 14, "Other Types of Vacuum Equipment").

Sometimes the vacuum was excellent, but only too often, the vacuum would break, and the asphalt would be contaminated with lighter and more volatile gas oil components. Mr. Wyatt phoned me at my home in New Orleans.

"Lieberman. I need you and your competent young wife, Liz, to come to Lake Charles. And I mean pronto."

"Mr. Wyatt, I can't. Not right away. I've got a job in Houston on Monday for the Crown Central Refinery. Maybe next month," I explained.

"No, Lieberman. I want you and Liz in Lake Charles tomorrow. Look, I'll send my private jet to collect you from New Orleans. After you fix the problem, the pilot will take you on to wherever you need to go. The jet will be waiting for you tomorrow morning."

"At what time Mr. Wyatt?"

"At whatever time you get there."

Let me tell you something, Ladies and Gentlemen. That's the only way to travel. By private jet. The copilot served us champagne and tiny sandwiches. We had our own private toilet. And when we flew on to Houston that afternoon, our plane was waiting for us – and only us.

The actual problem and solution were pretty simple. Let's refer to Figure 16.3. The problem originated when the feed rate to the vacuum tower was reduced slightly, or the vacuum tower flash zone temperature dropped by 10°F to 20°F, or the vacuum slipped a little. Any of these three factors would lead to a relatively minor loss in the vapor flow through the LVGO draw-off tray. The mechanical design of this tray was not the best. It was a bolted-up chimney tray rather than a welded chimney tray. Apparently it was subjected to excessive rates of LVGO product leakage at lower vapor flow rates. That is, whenever the pressure drop of the vapor flowing through the chimneys was appreciably reduced, the liquid level on the tray would be lost due to leakage. And then the following sequence of events would be precipitated:

1. The LVGO pumparound pump would lose suction and cavitate.
2. The vacuum tower top temperature would increase due to the reduction in the LVGO pumparound heat extraction.
3. The volume and weight of the vapor flow to the liquid ring seal pump would increase and the pump would bog down due to the increased vapor load.
4. The vacuum tower top pressure would increase, which suppressed the vapor flow to the LVGO draw-off pan, which then caused the LVGO chimney tray to leak worse. This caused even less liquid flow to the LVGO pumparound pump, which cavitated worse. The problem then built upon itself.

(Note – Process wise, the reader will recall, this is pretty much the same problem I described in the Delaware City Refinery, in Chapter 3.)

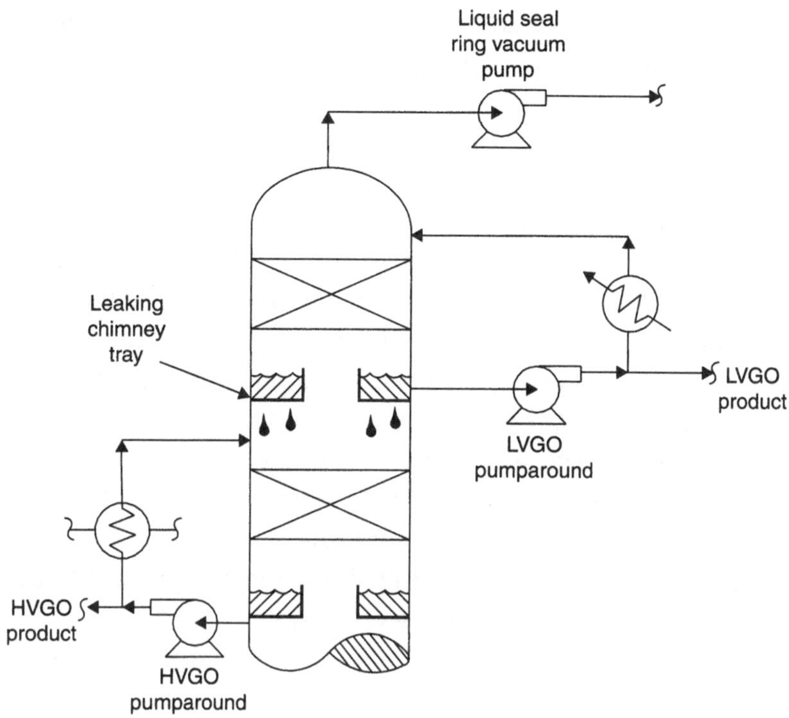

Figure 16.3 leaking LVGO chimney tray causes a sudden loss in vacuum.

The operators had become trapped in what I call a "Positive Feedback Loop". I've written an entire book on this subject, *Troubleshooting Process Plant Control*. But in this case, how were the Pelican Refinery operators to escape from this trap? How could they break free of this positive feedback loop?

I explained the following procedure to the plant operators:

"Look guys, when you see the net LVGO product LRC valve closing (see Figure 16.3), don't wait. You've got to take corrective action in just a few minutes, before the LVGO pump slips."

"Yeah, Mr. Norm. We don wanna lose that pan LVGO level. Then all hell breaks loose. But we don't got any more crude to run-off, and we're maxed out on the vacuum heater outlet temperature. What we all suppose to do then?"

"Okay," I explained, "But you all have got to act before the LVGO pan level falls. You need to reduce the HVGO pumparound rate by say 20%–30%."

"You see, gentlemen," Liz added, "Reducing the heat extracted in the HVGO pumparound will increase the vapor flow to the LVGO chimney tray. This will restore the vapor delta P through the chimneys and retard the rate of tray deck leakage. Also, more barrels of gas oil will condense out on the LVGO draw-off chimney tray, and also help in restoring the pan level."

"Right, Miss. Liz, we all got it. Makes good sense. Let's try it out right now," said Mr. Tiger Falcon, the chief operator. And they did, and it worked just fine.

I glanced at my watch, "Liz, it's time to go." Back at the airport, our jet and crew were waiting.

"We've filed our flight plan to Houston Hobby, Mr. Lieberman. Is that still correct?" asked the pilot.

"Yes Captain. But, Liz and I haven't dined yet."

"Oh, of course, sir. Henry will be serving a light dinner and a red wine onboard. A 1984 Merlot I believe."

And off we flew to Houston. But there was only one minor problem, which I really should not even mention. In my entire 29 years as a consultant, this is the only job I've done for free. I've only made note of this, on the off-chance that Mr. Oscar Wyatt were to read this story and wish to rectify his omission.

Improving Vacuum Tower Flash Zone Pressure

- Increase the bottom's stripping steam rate to the upstream crude tower. A poorly steam-stripped vacuum tower feed will overload the vacuum tower overhead ejector system.

- Minimize the operating pressure of the crude tower. This will also reduce the light ends in the vacuum tower feed. These light hydrocarbons will also contribute to overloading the vacuum jet system.
- Optimize the vacuum tower top pumparound circulation rate. If the vacuum tower has an overhead pre-condenser, you will find that increasing the top temperature often lowers the tower top pressure. The light ends will get "sponged up" in the heavy naphtha distilled overhead.
- Reduce the motive steam pressure to the jets. I once throttled back on the steam pressure to the second stage jets on a vacuum tower. The steam pressure was reduced from 160 psig to 115 psig. This improved the vacuum by only one mm of Hg, but saved 25% of the motive steam to these jets. A further reduction in the steam pressure, however, caused a loss in the vacuum.
- Back-flush the ejector condensers. Reducing the operating temperature of the condensers will always improve the vacuum to some extent. Acid cleaning of the condenser tube side will also help, as will blowing (air rumbling) the tubes out with plant air or nitrogen once a day.
- Check the motive steam moisture content. Wet steam will degrade the ejector performance and may actually cause a jet's nozzle periodically to freeze.
- Check for air leaks. Does the vacuum tower off-gas have more than 20% nitrogen? If so, there is a major air leak in the vacuum tower

(*Continued*)

(Continued)

or the vacuum system. An air leak into a vacuum system can be seen. When air expands into an area of lower pressure, the air is chilled. The resulting area of low temperature will cause moisture to condense around the leaking pipe or flange. Note that if the off-gas has very little O_2 but quite a bit of CO_2, the air leak is likely to be found in a hot area of the tower, not in the overhead system piping. By a "hot area", I mean the furnace transfer line in the region of high velocities.

- Listen carefully to the steam jets. Perhaps they are making a surging or "hunting" sound. This will lead to a big loss in vacuum. Often, this surging sound is caused by:
  - Overloading the jets with cracked gas.
  - Erosion of the steam nozzles.
  - High jet discharge pressure.
  - Wet steam.
  - Low motive pressure steam.
  - Air leaks.

16.3 Carry-Over of LVGO Pumparound Spray

At the now defunct Pacific Refining near Berkeley, California, I had made the following observations:

- The vacuum tower was a "Dry Tower", in that there was no pre-condenser.
- As I increased the top LVGO pumparound rate, the tower top temperature declined, which is normal.
- As the tower top temperature dropped, the vapor flow from the seal drum went down slightly,

which is fine. But the amount of hydrocarbon condensate from the seal drum increased, which was unexpected.

- The vacuum became slightly worse, which was also unexpected.
- The color of the condensate from the seal drum turned progressively more yellow (like the LVGO color) as the pumparound rate increased.

I then reasoned that the LVGO pumparound was misting up in the spray nozzles that were used to distribute the LVGO pumparound. This happens when one inch or smaller nozzles develop a delta P of more than about 40 psi. (Depends strongly on the type of nozzle and the manufacturer.) The LVGO mist was putting a small extra load on the first stage ejector.

To correct the problem and test my theory, I lowered both the pumparound return temperature and flow rate. My objective being to maintain a constant vacuum tower top temperature. The effects of this change were apparent in twenty minutes:

- The vacuum improved slightly.
- The yellow color of the seal drum condensate disappeared.

Note that this vacuum tower did not have a top demister. But demister corrosion, fouling and plugging only too often create a bigger problem than it solves. Therefore, I do not intend this story to be used as a justification for demisters in vacuum tower service. (Refer to the story later in this chapter.)

16.4 Optimizing Vacuum Tower Top Temperature

One would think that the optimum vacuum tower top temperature is always as low as possible. That is, to minimize the vapor load to the vacuum overhead system. However, as with everything else in vacuum systems, the situation is far from being that simple.

To start with, let's consider a dry vacuum tower. Meaning the vacuum tower overhead vapor flows first to the first stage ejector. Minimizing the tower top temperature certainly reduces the pounds of vapor flowing to the jet. However, if the tower top temperature gets too cold (i.e., below the water dew-point temperature), then water will begin to condense out inside the vacuum tower. The water will absorb the hydrochloric acid from the up-flowing vapors and cause horrendous rates of corrosion. The HCl is evolved from the residual salts in the crude oil that slip past the feed desalter.

But sometimes, I've observed that raising the tower top temperature reduces the suction pressure to the first stage jet, even with a dry vacuum tower. It's kind of complicated:

- Step 1 – Raising the tower top temperature increases the pounds of naphtha distilled overhead. The first stage ejector has excess capacity and is not significantly affected by the extra vapor flow.
- Step 2 – The extra naphtha condenses in the first stage condenser, which also has plenty of capacity to accommodate the extra duty.
- Step 3 – The naphtha sponges up some of the butane and pentane boiling range components from the gas flowing to the second stage ejector, which is, in this case, the limiting component in the vacuum system.
- Step 4 – The now unloaded second stage ejector reduces the discharge pressure of the primary ejector, which can then develop a reduced suction pressure, and hence a lower vacuum tower pressure.

For wet vacuum towers, meaning there is a pre-condenser ahead of the first ejector, increasing the tower top temperature often will improve the vacuum up to a point. Again, an increment of heavy naphtha is distilled overhead. This will absorb some of the butane boiling range components from the vapor flowing to the downstream ejector. The unloaded jet will then pull a better vacuum. Of course, at some point, the pre-condenser vapor outlet temperature will increase, and vacuum will be lost due to overloading the first jet.

The above effect, which I've seen often, gave me an idea, which I have never tried, but I'm sure it will work. Try pumping a small amount of stripped naphtha into the pre-condenser to act as a sponge oil. I recall that one of my clients has tried this technique, and they did report some improvement. However, they did not quantify the reduction in the vacuum tower top pressure when I asked them about the results of this modification.

16.5 Plugged Vacuum Tower Top Demister

The Texaco plant in Convent, Louisiana (now Motiva) has a vacuum tower with two large vacuum jets on top of the tower. I call this a dry vacuum tower. Even if velocity (i.e., heater pass coil steam) and bottom's stripping steam is used in the tower, it is still called a dry vacuum tower. If a pre-condenser is used, the tower is called a wet vacuum tower, even if no steam is employed in the vacuum tower.

In a dry vacuum tower, many process designers will place a demister on top of the vacuum tower—the idea being to reduce carry-over of entrained liquids into the primary jets, which could overload the jets. For myself, I'm not sure this does any good. My opinion is that you're better off without such a demister.

At the Texaco plant, I was troubleshooting a problem of excessive vacuum tower flash zone pressure. My field pressure survey showed the total delta P across the tower was only eight to ten mm of Hg, which is entirely normal. Basically, the problem was excessive tower top pressure. However, this was not due to the performance of the vacuum jets. There was a delta P of about 10–15 mm of Hg across the top demister.

These demisters look very much like a coarse, six-inch thick brillo pad. Their function is to coalesce the entrained droplets of gas oil into larger drops that will settle more readily, in accordance with Stoke's Law.

The vacuum tower was scheduled to be shut down in four months. I was supposed to produce a design to reduce the tower's flash zone pressure. The inlet pressure to the primary jets were six to eight mm of Hg. I was not likely to improve on this jet performance. The tower delta P across the three packed beds at eight to ten mm of Hg, as I mentioned above, was not likely to be reduced. So that left the 10–15 mm of Hg demister delta P. The calculated delta P across the demister was one mm of Hg, or 10% of the observed current delta P. This meant the damn demister was fouled, either with corrosion products, ammonia salts (NH_4Cl and $NH_4(HS)_2$), or both. So, I had two choices:

- Replace the demister

Or

- Discard the demister

What to do? I had to make a decision, but on what basis? I stood there in the blazing afternoon sun, melting in the August heat and humidity of Southern Louisiana. The Texaco operator, who had been assisting me in my pressure survey, offered me a pinch of Copenhagen (chewing tobacco).

"No thanks, Jeff," I said.

"You all know, Mr. Norm, somethin' strange happened on this here tower a coupla years ago."

"What was that, Jeff?" I asked.

"Well, I'll tell you," Jeff said after spitting tobacco juice on a steam trap. "The vacuum started to get worse 'n worse. Pressure in the flash started to build. Over a year, crept up by maybe twenty millimeter."

"And then?" I asked.

"And then, Mr. Norm, one day I's sitting there minding the panel like always, and the flash zone pressure drops sudden like. Somethin' like twenty millimeter in minutes."

"Twenty millimeters of mercury in just a few minutes?"

"Yeah. And then everything lined out jus' fine. Gas oil yield increased and vacuum bottom's flow tailed-off."

"And then?" I asked.

"And then nothin'. We just ran like that for the rest of the year. Til the next turnaround. When we opened the vacuum tower, I seen that demister pad. You all know at the top of the tower. It don tore itself away from its support. It was still there, but half of it was pushed up and off to one side. Mr. Norm, you all think them demisters really does us any good?"

"I guess not, Jeff. Maybe we'll leave it out of this tower."

"Yeah, Mr. Norm. I mentioned all this last year to Mr. Baubien, but he said we need the demister to get a good vacuum. To protect the big jets atop the vacuum tower from entrainment droplets type-a-stuff. But that don't seem that-a-way to me."

"I agree with you. But Jeff, did the vacuum condensate liquid flow from the seal drum increase after you blew out the demister?"

"Nope. Stayed bout the same. I told Mr. Baubien that too. But he never paid no mind to us operators," Jeff said as he spit on a pump. "He always had some portant meetin to go to. Mr. Norm. Sure you don't want just a pinch?" Jeff asked, as his spit sizzled away on the hot pump case. "I got me a fresh can."

16.6 Bypassing Primary Ejector

Last week I had an interesting experience at a 30,000 BSD feed vacuum tower. The first stage jet inlet pressure was 30 mm Hg and the downstream first stage condenser outlet pressure was 45 mm Hg. To optimize the motive steam pressure to the first stage jet, I slowly throttled the steam supply valve. Both the inlet pressures to the first jet and outlet pressure from the first condenser dropped. When I had shut off the steam totally to the primary jet, its inlet pressure had dropped to 28 mm Hg. The outlet pressure from the condenser had also fallen, to 24 mm Hg. Next, I opened the bypass line around the now idled jet and its inlet pressure fell to a satisfying 26 mm Hg.

Not only had I reduced the vacuum tower pressure by four mm Hg (from 30 mm to 26 mm Hg), but I had also saved the 6,000 lbs per hr of 180 psig motive steam.

In this case, the jet was not developing much of a compression ratio (i.e., 45 mm ÷ 30 mm = 1.5). The downstream condenser was heavily loaded from the motive steam flow to the upstream jet. Then, reducing the steam flow to the jet unloaded the condenser and reduced its condensing temperature and condensing pressure. This reduced condensing pressure more than offset the loss of the small 1.5 compression ratio produced by the first stage steam ejector.

In this case, the jet was overloaded because only one of the two parallel jets was in service. The limited first stage condenser capacity precluded the use of motive steam to both jets simultaneously. The malfunction with the first stage condenser was a design error in the placement of the air or vapor baffle too close to the non-condensible outlet nozzle, which restricted the flow of gas to the vapor outlet nozzle.

17

Frequently Asked Questions

I'll spend several hours each week answering questions from readers and former students that I receive via email. The most common subject for these questions pertains to motor driven (constant speed) centrifugal compressors. The second most common subject for questions is concerned with converging-diverging steam ejectors and their effect on vacuum tower performance. I've attached in this chapter, a sampling of the most commonly asked questions (edited for clarity). Please feel free to forward any of your own questions to me at norm@lieberman-eng.com.

17.1 Vacuum Systems

Dear Mr. Norman – Our vacuum tower top pressure is 40–45 mm of mercury. We have two large vacuum jets on top of the tower. The front part of both jet bodies is around 80°F.

The steam supply is 150 psig and 360°F. The temperatures are obtained using my infrared gun. Is that 80°F normal, good, or bad?
– Henry R.

Henry – The 150 psig, 360°F motive steam may be saturated, or it may be wet. When steam expands through the ejector nozzle, its temperature is converted to velocity. The more efficient the conversion, the higher the velocity of the steam. The greater the velocity of the steam, the more efficient the compression of the vacuum tower off-gas. Hence, in that sense the cooler the mixing chamber (i.e., the front end of the jet), the better.

However, if the mixing chamber is cool, that could also be due to moisture in the supply steam, flashing as it enters the mixing chamber. This conversion of latent heat to sensible heat just robs energy from the motive steam, and thus slows down the motive steam as it enters the diffuser. And this is bad.

If the part of the diffuser immediately downstream of the mixing chamber doesn't get too hot to touch within one or two feet of the mixing chamber, that's an indication of wet steam.
–Norm

Mr. Norman Lieberman: I attended your Troubleshooting Seminar in 2001 in Ft. McMurray. I have a question relating to my vacuum tower (I'm now working in a refinery on the Gulf Coast). What does it mean if ice forms on the outside of the ejector body?

Thank you for your help. I really enjoyed your seminar.
–Frank.

Frank – This is a sure indication of water in the motive steam supply. Not just a little water, but probably over 10%. If you can eliminate this moisture, I'm sure that you will see an immediate improvement in your vacuum tower performance.

–Norm

Norm – I was reading *Troubleshooting Process Operations*. In your chapter on vacuum towers, you talked about gas composition of the seal drum off-gas. I sampled this stream. It is 40% hydrogen sulfide. Is this possible? My boss said that I have done something wrong.
–Raymond

Raymond – First, let me warn you to be careful. One breath of that off-gas will knock you out. Yes, I've measured up to 40% myself in Aruba, when the vacuum tower bottom's was about 4% sulfur. There is not much to be done to reduce the H_2S concentration. More typically, I see off-gas in the 10%–20% H_2S range; this will knock you out the same as 40%. I hope this helps with your supervisor.
–Norm

Dear Dr. Lieberman – I have purchased many of your books in which you have kindly invited questions. My question pertains to our new vacuum tower operation. Every morning about 10:00 a.m., our vacuum ejectors start to make a regular sound. The operators say the ejectors start to breathe. At the same time, our flash zone pressure increases from 10 to 12 mm of mercury to about 20. Any suggestions you may offer would be sincerely appreciated.
–Amed

Amed – Your jets are surging. Most likely, that's because of higher discharge pressure. The cause of the higher discharge pressure is probably an increase in the plant's cooling water temperature as the sun comes up. When I say your jets are surging, I mean that they are losing their sonic boost as described in my book, *Process Equipment Malfunctions*.

To start with, try back-flushing the condenser water side on the discharge of the surging jet. Unfortunately, there are dozens of different things that can cause jets to surge. But anything

you can do to improve water flow through the downstream condenser is sure to help to some degree.
–Mr. (not doctor) Norman Lieberman

Mr. Lieberman – Which should be hotter – the vapor or liquid outlet of my vacuum tower pre-condenser? Currently, the liquid outlet is approximately 20°C and the vapor outlet is 40°C. That difference is increasing and our tower top vacuum is becoming progressively worse.
–L.T. Lee

To Mr. L.T. Lee – By design, the vapor outlet of a normal surface condenser (where the vapor outlet is located on the side of the condenser shell) is cooler by about 20°F–30°F. This is accomplished by an internal baffle located inside the tube bundle. It's called the air or vapor baffle. It forces the vapor to flow down and then up across the tubes.

There are two possible reasons for your problem. First, this air baffle may be leaking. This is always caused by defective air baffle seal strips (Please see my book, *Process Equipment Malfunctions*, McGraw Hill, 2011.)

Secondly, there may be condensate back-up due to inadequate drainage from the seal leg. Your seal leg may be plugging, or your seal drum may be filling with corrosion products or sludge. Blowing out the seal legs with steam will help in this case. Perhaps the seal leg is sucking in air due to a seal leg leak. Such a leak will prevent proper drainage through the seal leg. Check for this by raising the liquid level in the seal drum.

Regardless of the cause, an increase of the vapor outlet temperature is certain to cause a loss of vacuum and could possibly result in the jets making a surging (i.e., erratic) sound, which is a certain indication of a loss in the sonic boost of the jets.
–Norm

To Norman Lieberman: Norman. We met in India last year. I have a serious problem. Our vacuum on our asphalt vacuum

tower is always bad. But sometimes it suddenly becomes much worse. It seems to correlate with increased naphtha product from the seal drum of the vacuum tower overhead system. As a result, we cannot produce proper grade of asphalt for sales. Any advice would be greatly appreciated.
–Kumar

Kumar – If the naphtha production rate is very high, the naphtha liquid level in the first stage condenser (or pre-condenser) may start to back-up. This reduces the surface area exposed to the condensing vapor, which consequently will increase the vapor load to the downstream ejector. However, if the naphtha liquid level rises to the bottom edge of the air or vapor baffle inside the condenser shell, then the vapor is trapped inside the condenser. Pressure will build inside the condenser shell until the pressure is great enough to push the liquid level down below the bottom edge of the air baffle, so that the vapor can escape out of the condenser.

On one occasion, I found this problem to be caused by a high liquid level in the seal drum, which was a consequence of an erroneous seal drum level indication. Also, the seal drum naphtha pump had lost capacity due to a worn pump impeller wear ring. Hope this helps.
–Norm

Mr. Lieberman – I have a vacuum tower question. Reducing the ejector steam supply below its design (vendor specified) pressure improves vacuum. How could this happen? I suppose it's best to operate at the lower seven bar pressure rather than the design ten bar of steam pressure.
–Sriganesh

Sriganesh – Likely one third of the jets I look at have the property that reducing the motive steam supply pressure below the design value improves vacuum, or at least does not harm the vacuum.

The two most common reasons for this are that the ejector steam nozzle is badly worn, or that the downstream

condenser is overloaded. Either way, it's an indication of poor system performance.
–Norm

Dear Mr. Norman P. Lieberman – I was reading your chapter on vacuum system operation. You talked about how raising the level in the seal drum can improve vacuum. I have observed this on our vacuum tower myself. Certainly, this is not normal. We have to flood our seal drum to sustain a reasonable vacuum.
–Harvey C.

Harvey – Unfortunately, you have developed a leak in the barometric drain line (i.e., your seal leg inside the seal drum). This is caused by biological corrosion or a leaking internal flange. Raising the level to submerge this leak stops the seal drum gas from getting sucked-up into the leg and interfering with this drainage. I know it seems backwards, but flooding the drum will lower the condensate back-up level in the condenser. This lowers the vapor outlet temperature from the condenser and thus unloads the downstream ejectors.

I suggest that you add some biocide to your seal drum to suppress the biological corrosion. Then, during the next turnaround, replace the carbon steel seal legs with 316 (L) S.S., or whatever alloy is suitable for your service.

I've discussed a similar incident in my book, *Troubleshooting Process Operations, 4th Edition*, PennWell Publications.
–Norm

Mr. Norman Lieberman – What does it indicate if we have a gradual increase in CO_2 in the vacuum tower waste gas to the vacuum heater? A year ago the CO_2 concentration was one percent and now it is ten percent. Our vacuum is still quite acceptable. Also, the nitrogen (but not oxygen) concentration of the waste gas stream is increasing, which I assume indicates an air leak.
–Carl

Carl – You most likely have an air leak in the vacuum heater transfer line feeding the vacuum tower. The leak is getting progressively worse. Since the leaking line is under a slight vacuum, relatively small amounts of air are drawn into the vacuum tower. The oxygen then reacts with the hydrocarbons to form CO_2 (and perhaps some CO). Regardless, if you have a sudden loss of vacuum, the area of the leak will pressure-up. Hot vacuum tower feed will erupt from such a leak and auto ignite. This is a very dangerous situation, and I've been involved in such incidents three times in my career.

I don't really know how to advise you. Shutting down to repair the leak in the transfer line may very well result in a fire. One of my clients burned down their vacuum tower due to such a failure. Sorry for the bad news, but I believe my analysis is quite accurate.

–Norm

Mr. Norman P. Lieberman – Why does an increase in the tower bottom's level in my vacuum tower cause a loss in vacuum? Does this indicate an excessively high tower bottom's level?

–George

George – No. The level is not the problem. It's that your boot temperature is too hot. The problem is excessive cracked gas formation in the boot. Assuming your level change is within the level span of the boot, increase the level from 30% to 70%. If the vacuum is adversely affected, try lowering the vacuum tower bottom's temperature by 10°F, using the circulating boot quench. You are also promoting coke formation in the boot, which may coke-up the suction of your vacuum tower bottom's pumps.

–Norm

Mr. Norman Lieberman – We have read, and noticed ourselves, that if there is too much superheat in the motive steam, vacuum is lost. But also, if the steam pressure gets too high, vacuum also gets poor. We think that the more steam pressure and the more superheat, the more energy of the motive steam. The more energy of the motive steam, the harder and better the steam ejector should work and the bigger its compression ratio.

Hope you can help us.

–Ashok

Dear Ashok – Regarding superheated steam, let's say I increase the temperature of 150 psig steam from 400°F to 600°F. From my steam tables or Mollier Diagram, I see that:

- The enthalpy of the steam increases by 100 BTU/LB, or perhaps ten percent, which is good.
- Then density of the steam decreases, due to the higher absolute temperature (860°R to 1060°R) by about twenty-two percent, which is bad.

As the flow of steam is limited by sonic velocity in the steam nozzle, superheating the steam by the 200°F would result in:

$$+10\% - 22\% = -12\%$$

That is, decrease the amount of steam energy to the ejector by twelve percent.

Regarding, increasing steam pressure: In the field, I often find that a higher steam pressure improves the vacuum, unless the downstream condenser is limiting. Unfortunately, the downstream condenser is often limiting.

For instance, increasing the motive steam pressure from 150 psig to 180 psig, increases the steam density by about eighteen percent. This would increase the velocity of the steam in the ejector's diffuser by roughly the same percentage. But, the downstream condenser heat duty would increase by perhaps ten percent.

This is bound to raise the condensing pressure by some amount, depending on the condenser's excess capacity. The resulting back-pressure at the discharge of the ejector, will bog the ejector down and may lead to the loss of its sonic boost.

I hope this helps. Best Regards.

–Norm

Mr. Lieberman – Can you suggest an effective method to clean the shell-side of our badly fouled first stage condenser on-line, without shutting down? The delta P across the shell-side is two psi—more than ten times the design. It's all a single train system and we are not able to shut down our vacuum tower for another two years.

–John Q.

Dear John – I've investigated your problem recently for several clients. I am unaware of any on-line procedure or chemical that would even partially alleviate your very common problem of excessive condenser pressure drop. Best to bypass the condenser, shut off the steam to the upstream jet, and chemically clean the condenser. The tower pressure may not increase to an unacceptable level.

–Norm

You may contact me if you have any questions pertaining to your vacuum system problems at:

- 1-504-887-7714 (phone)
- 1-504-456-1835 (fax)
- norm@lieberman-eng.com
- www.lieberman-eng.com

The Norm Lieberman DVD/Video Library of Troubleshooting Process Operations

Intended for experienced operating personnel & plant process engineers. Each DVD runs 35 to 55 minutes.

TO ORDER:

- Consult our website: http://www.lieberman-eng.com/
- E-mail us at norm@lieberman-eng.com
- Phone: 1-504-887-7714
- Fax: 1-504-456-1835

Each order of ten or more DVD's receives a complete complementary set of Lieberman text books.
Please specify country format required for DVD's.

VACUUM SYSTEMS & CONDENSERS:

- Steam Ejectors
- Vacuum Tower Overhead Systems
- Surface Condensers
- Subcooling Robs Condenser Capacity

- Fundamentals of Condensation
- Back Flushing & Acid Cleaning Surface Condensers
- Pressure Control of Columns
- Improving Vacuum Tower Operation
- Vacuum Towers – Packing & Grids

CENTRIFUGAL PUMPS:

- Fundamentals of Centrifugal Pumps
- Optimizing Impeller Size
- Cavitation & Net Positive Suction Head
- Causes of Seal & Bearing Failures

COMPRESSION:

- Fundamentals of Gas Compression
- Compressor Surge
- Variable Speed Centrifugal Compressors
- Steam Turbine Drives
- Amp Load on Motor Driven Compressors
- Reciprocating Compressors

DISTILLATION:

- Distillation Fundamentals
- Jet Flood & Dry Tray Pressure Drop
- Incipient Flood – A Basic Concept
- Identifying Flooding Due to Foaming
- Tray Dumping & Weeping
- Tower Pressure Drop Evaluation
- Plugging & Fouling of Trays
- Packed Tower Fundamentals
- Coker, FCU & Crude Fractionators

- On-Stream Cleaning of Trays
- Structured Packing
- Optimizing Pumparounds
- Steam Strippers
- Inspecting Tower Internals

HEAT EXCHANGERS & REBOILERS:

- Heat Transfer Fundamentals
- Once-Through Thermosyphon Reboilers
- Circulating Thermosyphon Reboilers
- Kettle Reboilers
- Forced Circulation Reboilers
- Foam Formation in Reboilers
- Steam & Condensate Flow
- Water Hammer – Causes & Cures
- Shell & Tube-Exchanger Fouling

FIRED HEATERS & DRAFT:

- Fired Heater Fundamentals
- Optimizing Excess Air
- Draft Balancing Heaters
- Air Preheater Problems
- Preventing Heater Tube Coking & Fouling

TREATING & SULFUR RECOVERY:

- Amine Regeneration & Scrubbing
- Amine Degradation & Foaming
- Sulfur Plant Start-up Problems
- Sulfur Plant Pressure Drop
- Kerosene or Jet Fuel Treating

SAFETY UPGRADING:

- Field Testing Alarms & Trips
- Failure of an Amine C_3—C_4 Scrubber
- Boiling Water – A Safety Hazard
- Sulfur Plant Safety Hazards
- Sources of Auto-Ignition

REFINERY PROCESSES:

- Sulfuric Acid Alkylation
- Crude Tower Overhead Corrosion
- HF Alkylation Pressure Control Problems
- Delayed Cokers – Decoking cycle
- Delayed Cokers – Fractionation

Index

Also of Interest

Check out these other related titles from Scrivener Publishing

Bioremediation of Petroleum and Petroleum Products, by James Speight and Karuna Arjoon, ISBN 9780470938492. With petroleum-related spills, explosions, and health issues in the headlines almost every day, the issue of remediation of petroleum and petroleum products is taking on increasing importance, for the survival of our environment, our planet, and our future. This book is the first of its kind to explore this difficult issue from an engineering and scientific point of view and offer solutions and reasonable courses of action. *AVAILABLE SEPTEMBER 2012.*

Sustainable Resource Development, by Gary Zatzman, ISBN 9781118290392. Taking a new, fresh look at how the energy industry and we, as a planet, are developing our energy resources, this book looks at what is right and wrong about energy resource development. *AVAILABLE NOVEMBER 2012.*

An Introduction to Petroleum Technology, Economics, and Politics, by James Speight, ISBN 9781118012994. The perfect primer for anyone wishing to learn about the petroleum industry, for the layperson or the engineer. *NOW AVAILABLE!*

Ethics in Engineering, by James Speight and Russell Foote, ISBN 9780470626023. Covers the most thought-provoking ethical questions in engineering. *NOW AVAILABLE!*

Fundamentals of the Petrophysics of Oil and Gas Reservoirs, by Buryakovsky, Chilingar, Rieke, and Shin. ISBN 9781118344477. The most comprehensive book ever written on the basics of petrophysics for oil and gas reservoirs. *NOW AVAILABLE!*

Petroleum Accumulation Zones on Continental Margins, by Grigorenko, Chilingar, Sobolev, Andiyeva, and Zhukova. Coming in September 2012, ISBN 9781118385074. Some of the best-known petroleum engineers in the world have come together to produce one of the first comprehensive publications on the detailed (zonal) forecast of offshore petroleum potential, a must-have for any petroleum engineer or engineering student.

Mechanics of Fluid Flow, by Basniev, Dmitriev, and Chilingar. Coming in September 2012, ISBN 9781118385067. The mechanics of fluid flow is one of the most important fundamental engineering disciplines explaining both natural phenomena and human-induced processes. A group of some of the best-known petroleum engineers in the world give a thorough understanding of this important discipline, central to the operations of the oil and gas industry.

Zero-Waste Engineering, by Rafiqul Islam, ISBN 9780470626047. In this controversial new volume, the author explores the question of zero-waste engineering and how it can be done, efficiently and profitably. *NOW AVAILABLE!*

Formulas and Calculations for Drilling Engineers, by Robello Samuel, ISBN 9780470625996. The most comprehensive coverage of solutions for daily drilling problems ever published. *NOW AVAILABLE!*

Emergency Response Management for Offshore Oil Spills, by Nicholas P. Cheremisinoff, PhD, and Anton Davletshin, ISBN 9780470927120. The first book to examine the Deepwater Horizon disaster and offer processes for safety and environmental protection. *NOW AVAILABLE!*

Advanced Petroleum Reservoir Simulation, by M.R. Islam, S.H. Mousavizadegan, Shabbir Mustafiz, and Jamal H. Abou-Kassem, ISBN 9780470625811. The state of the art in petroleum reservoir simulation. *NOW AVAILABLE!*

Energy Storage: A New Approach, by Ralph Zito, ISBN 9780470625910. Exploring the potential of reversible concentrations cells, the author of this groundbreaking volume reveals new technologies to solve the global crisis of energy storage. *NOW AVAILABLE!*